Restraint and Handling for Veterinary Technicians and Assistants

Join us on the web at

agriculture.delmar.com

Restraint and Handling for Veterinary Technicians and Assistants

Bonnie Ballard, DVM

Jody Rockett, DVM

DELMAR
CENGAGE Learning™

Australia • Brazil • Japan • Korea • Mexico • Singapore • Spain • United Kingdom • United States

Restraint and Handling for Veterinary Technicians and Assistants

Bonnie Ballard, DVM
Jody Rockett, DVM

Vice President, Career and Professional Editorial: Dave Garza

Director of Learning Solutions: Matthew Kane

Acquisitions Editor: Benjamin Penner

Managing Editor: Marah Bellegarde

Senior Product Manager: Darcy M. Scelsi

Editorial Assistant: Scott Royael

Vice President, Career and Professional Marketing: Jennifer McAvey

Marketing Director: Deborah Yarnell

Marketing Manager: Erin Brennan

Marketing Coordinator: Jonathan Sheehan

Production Director: Carolyn Miller

Production Manager: Andrew Crouth

Content Project Manager: Katie Wachtl

Senior Art Director: Dave Arsenault

All figures Delmar/Cengage Learning

Library of Congress Control Number: 2009921511

ISBN-13: 978-1-4354-5358-6

ISBN-10: 1-4354-5358-1

Delmar
5 Maxwell Drive
Clifton Park, NY 12065-2919
USA

Cengage Learning is a leading provider of customized learning solutions with office locations around the globe, including Singapore, the United Kingdom, Australia, Mexico, Brazil, and Japan. Locate your local office at: **www.cengage.com/global.**

Cengage Learning products are represented in Canada by Nelson Education, Ltd.

To learn more about Delmar, visit **www.cengage.com/delmar**

Visit our corporate website at **www.cengage.com.**

Notice to the Reader

Table of Contents

Chapter 4.   Restraint of Cats   52

Chapter 5.   Restraint of Exotic Pets   69

Unit 2: Large Animal Restraint

Preface

Techniques of restraint are applied in the everyday working lives of veterinary technicians and DVMs. Great care must be taken to ensure the safety of the animals being cared for as well as to maintain your safety and that of the coworkers you are handling the animals with.

The material in each chapter is presented in a uniform and easily understandable format. A deviation from the standard prose format has been adopted, intentionally, to provide concise answers to critical questions everyone has related to the techniques of restraint: "What do I need to do, and what can go wrong?" The goal of the book is to provide these answers in a clinically accessible format, eliminating the need to wade through more traditional texts.

With this goal in mind, the text is constructed such that the purpose for each procedure is clearly stated, followed by a list of potential complications. Equipment requirement lists precede the step-by-step instructions for performing each procedure. Finally, procedural instructions are augmented by short explanatory statements or comments.

Acknowledgments

I would like to thank Dr. Richard Duffey, owner of Winder Animal Hospital, for providing all the photographs of small animal restraint. Thanks also goes to Jody Woodard, RVT, Heather Bohmann, and Michele Godina, RVT for serving as restrainers for the pictures. Thank you to Lucy, Mr. Bigglesworth, Chatzee, Silver, Cage, Homer, Ozzie, and Cody for demonstrating restraint techniques.

I would also like to thank Kate Beckman, RVT, Daniel Vela and Thandi Powell from All Creatures Animal Hospital for their help with the exotic chapter photographs.

Lastly I would like to thank my husband Brian Kershaw for helping me edit the text and providing the encouragement I needed to complete this project.

Unit 1

Small Animal Restraint

Chapter 1

Principles of Restraint in Veterinary Practice

We can judge the heart of a man by his treatment of animals.

—Immanual Kant

Objectives

- Understand why restraint is necessary.
- Discuss why appropriate restraint techniques vary from species to species.
- Explain why different temperaments and situations require different restraint techniques.
- Explain the types of procedures for which proper restraint is required.
- Explain the importance of animal safety when using restraint techniques.

Key Term

restraint

Principles of Restraint

Restraint is defined as forcible confinement; in veterinary practice it is the forcible confinement of an animal within the practice's care. Veterinary technicians and assistants will be using restraint techniques every day in their professional lives. It is a skill that takes practice to master and to feel confident performing. The goal of restraint is to handle an animal in such a way that a procedure can be done without injuring the animal and without causing any injury to the humans involved in the procedure. It should always be the objective of the restrainer to never let the person performing the procedure get hurt. Anyone who is performing a procedure wants to be able to concentrate on the task at hand and not have to worry about what the animal is doing. A technician or assistant should have the attitude that "no one will get hurt while I am restraining the animal." The potential for serious injury is ever present.

Because the veterinary facility is liable for any injury an owner sustains from their pet, owners should not be allowed to perform restraint. It goes without saying that the average owner is not likely trained in proper restraint techniques.

Different animals require different restraint techniques. For example, cats usually do well with minimal restraint, whereas a dog typically requires more control. Individual animal temperaments will also dictate which technique is appropriate. A friendly dog may require less restraint than a fearful or aggressive dog.

Some techniques are appropriate to perform in front of the pet's owner, whereas others are not as they may look harsh. For example, the cat "stretch" restraint technique, while not harmful to the cat, may be viewed by the owner as cruel.

Different restraint techniques are required for the performance of veterinary procedures. For example, an unpleasant procedure such as a rectal examination will require more restraint than auscultation of the heart. Certain procedures require special techniques. For example, the restraint for obtaining blood from the jugular vein is vastly different than that required for venipuncture of the medial saphenous vein.

In all situations, regardless of species, one must have patience. If a restrainer is in a bad mood or is pressed for time, this could make a difficult situation worse.

Animals in a Veterinary Facility

An animal in a veterinary facility may not be the animal that its owner knows at home. The normally calm animal may be excited. An animal that never shows aggressive tendencies in its day-to-day life may be aggressive in the clinic. An animal may appear perfectly calm but then lash out when a clinic employee tries to handle it. Although the profession of veterinary medicine is chosen by individuals who love animals, caution must be taken whenever a strange animal is approached in the veterinary clinic. Much of the adverse behavior seen is born out of fear, strange surroundings, and unusual scents. With time and experience, veterinary personnel learn to "read" an animal's body language before approaching it in the clinic. Although not within the scope of this text, a study of the basic behavior of the animals to be worked with is essential to ensuring safe encounters with patients. Behavior resources are listed at the end of the chapter. The bottom line is that *all* animals should be approached with caution even if they appear friendly.

Precautions should be taken to make sure that the animal does not escape from the clinic if it escapes the restraint technique. If a procedure is being performed in a treatment room, all doors should be closed. This is also true if the procedure is performed in an examination room. Doors to rooms adjoining the one where the restraint is being performed should be closed as well.

If a technician or assistant gets in a situation where an animal cannot be safely restrained, help should be sought. Pride should never get in the way of personal safety. Be aware that certain animals in certain situations cannot be adequately restrained regardless of the level of experience the restrainer has. This is when the decision to use tranquilization or sedation comes into play.

Animal restraint is one of the most important skills a technician and assistant should possess. Being a master at restraint is a trait that will be valued in a veterinary clinic. Practice makes perfect!

Complications of Restraint

Restraint is required for proper transportation, examination, and treatment of any animal species. The degree of restraint required reflects the species, the animal's familiarity with handling, anticipated invasiveness, and the duration of the procedure. It is the handler's responsibility to use techniques that facilitate the success and safety of all humans and animals involved in a procedure. Unfortunately, despite all attempts to minimize complications, restraint can adversely affect some animals. Undesirable effects that can be associated with restraint include:

- Trauma, including contusions, bruising, lacerations, and nerve paralysis
- Metabolic disturbances such as acidosis, hypoxia, hypocalcemia, hyperglycemia, and hypoglycemia
- Hyperthermia
- Regurgitation
- Emotional stress

Review Questions

1. What is the definition of restraint?
2. What is the goal of restraint?
3. Why do restraint techniques vary by species?
4. What is the relationship between temperament and the restraint techniques used?
5. Under what circumstances is restraint required?
6. Why is safety important when using restraint techniques?
7. Why should an owner not be allowed to restrain their pet?
8. What can be done to prevent an animal from escaping if it gets away from a restrainer?
9. Why is it important to be able to read a pet's body language?
10. What are some of the adverse effects of restraint that could occur in animals?

Bibliography

Crow, S., & Walshaw, S. (1997). *Manual of clinical procedures in the dog, cat and rabbit*. Philadelphia: Lippencott-Raven.

Fowler, M. (2008). *Restraint and handling of wild and domestic animals* (3rd rev. ed.). Ames: Iowa State University Press.

McCurnin, D., & Bassert, J. (2006). *Clinical textbook for veterinary echnicians* (6th ed.). St. Louis: Elsevier Saunders.

Sheldon, C., Sonsthagen, T., & Topel, J. (2006). *Animal restraint for veterinary professionals* (2nd ed.). St. Louis: Mosby Elsevier.

Supplemental Reading

Landsberg, G., Hunthausen,W., & Ackerman, L. (2003). *Handbook of behavior problems of the dog and cat*. Philadelphia: Saunders.

Overall, K. (1997). *Clinical behavioral medicine for small animals*. St. Louis: Mosby.

Chapter 2

Restraint Tools and Techniques for Small Animals

The purity of a person's heart can be quickly measured by how they regard animals.

—Anonymous

Objectives

- Identify the types of restraint tools available and explain how to maintain them.
- Identify which situations require which tool.
- Discuss the proper way to apply restraint to avoid injury to the animal.

Key Terms

cat bag
muzzle
noose leash
rabies pole

The Use of Restraint Tools in Small Animal Practice

There are many tools at a veterinary technician or assistant's disposal to restrain animals. Some can be purchased through veterinary supply companies. Other restraint devices can be made from materials commonly found in a veterinary hospital. For example, a large towel can be used as a capture and/or restraint device when dealing with a cat or a small dog. A roll of gauze can be used to make a muzzle if a leather or nylon one is not available. Doors in the hospital can be used as a squeeze cage. While not featured in this chapter, there are many different types of devices that can be purchased such as the "clam shell", capture net, and grasper to catch and restrain cats. The most common restraint methods and devices used in veterinary practices are explained here.

The goal of using restraint devices is to make dealing with the animal safer for the restrainer and avoid injury to the animal being restrained. Interestingly, in many cases, the use of some restraint devices seems to cause the animal to "give up." For example, it is not uncommon to see a particularly fractious dog completely calm down once a muzzle is applied.

Noose Leashes

A **noose leash** is a single piece of nylon or rope with a ring on one end and a handle at the other. The handle end is fed through the ring to make the noose, which goes around the animal's neck. A noose leash is used to walk a dog or to remove a dog or cat from a run or cage. It is also used as a means of controlling an animal if it gets loose from a restrainer's hold. In most of the techniques described in Chapter 3, it is suggested that a noose leash be used for this purpose.

Using a Noose Leash to Remove a Dog from an Enclosure

The noose leash can be used to remove a dog from a cage or run. This is especially helpful if the veterinary technician is not sure that the animal is friendly enough to touch or if the dog is too large to lift out of the enclosure. Note that this technique can also be used to remove fractious cats from a cage.

SAFETY ALERT

- One must be careful not to choke an animal when using a noose leash.
- Nylon leashes should be washed regularly to lessen the spread of disease.
- An animal should never be left unattended or tied to something using the leash.

SAFETY ALERT

Any fractious animal should always be placed in the lowest cage possible. Removing an animal from a cage that is chest level is not only unnerving but also dangerous. At this level, the animal's mouth is in dangerous proximity to the technician's face!

PROCEDURE
Using a Noose Leash to Remove a Dog From an Enclosure

Technical Action

1. Obtain a noose leash.

Rationale/Amplification

1a. An example of a noose leash is shown in *Figure 2-1a*.

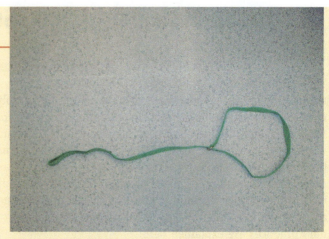

Figure 2-1a: Noose leash.

Technical Action

2. Make the loop on the noose leash big enough to fit around the head.

Rationale/Amplification

2a. You will need to "eyeball" this to guess the proper size. *See Figure 2-1b*.

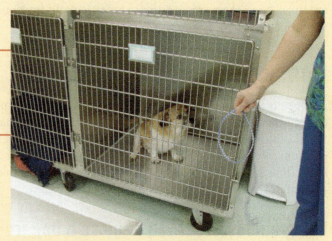

Figure 2-1b: Making a loop in the leash.

Technical Action

3. Open the cage door or run door just enough to slide the noose leash loop in and over the dog's head.

Rationale/Amplification

3a. You only want to allow for the leash to fit in and not the dog to escape. *See Figures 2-1c and 2-1d*.

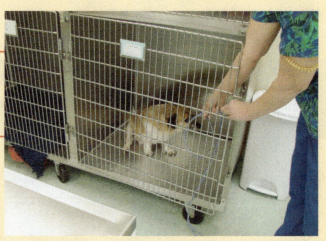

Figure 2-1c: Opening the cage to get the leash over the dog's head.

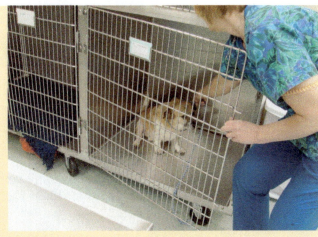

Figure 2-1d: Placing the leash around the dog's neck.

Technical Action

4. Once the dog's head is in the noose, tighten it and open the door, allowing the dog to exit the enclosure.

Rationale/Amplification

4a. *See Figure 2-1e.*

4b. Getting the dog's head in the noose may take a couple of attempts if the dog is fearful, aggressive, or uncooperative.

Figure 2-1e: Leading the dog out of the cage.

Returning a Dog to an Enclosure Using a Noose Leash

Once a procedure is finished, the dog will need to be returned to its cage or run. This may require that the dog be lifted into the cage if it won't walk in voluntarily.

PROCEDURE
Returning a Dog to an Enclosure Using a Noose Leash

Technical Action

1. Walk the dog to the cage or run.

Technical Action

2. Open the door and see if the dog will walk in voluntarily.

Rationale/Amplification

2a. *See Figure 2-2a.*

2b. If it will not, you may need to give a gentle verbal command to encourage it or lift the front limbs into the enclosure.

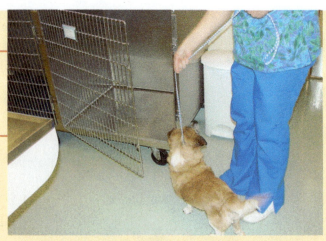

Figure 2-2a: Leading the dog into the cage.

Technical Action

3. Let the dog enter the cage, holding the leash with one hand.

Rationale/Amplification

3a. *See Figure 2-2b.*

Figure 2-2b: Encouraging the dog to step up into the cage.

Technical Action

4. Once the dog is in the cage, holding the door with one hand, use your other hand to remove the leash from the dog's head.

Rationale/Amplification

4a. The door should be held open only enough to get the leash off. If the door is opened too wide, the dog may try to escape. *See Figure 2-2c.*

4b. If the dog is unfriendly, you may need to try to loosen the leash so that the noose will come off without touching the dog.

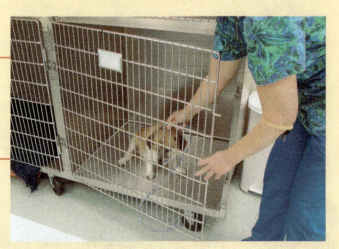

Figure 2-2c: Removing the leash.

Technique for Removing a Noose Leash from a Fractious Dog

This technique allows a restrainer to use a noose leash with another one attached, which is used to remove the leash when the dog is too aggressive to remove it with your hands.

PROCEDURE
Removing a Noose Leash From a Fractious Dog

Technical Action

1. You will need two noose leashes.

Rationale/Amplification

1a. The green leash will be the one that goes around the dog's neck. The orange leash is the one that will be attached for removal of the green leash. *See Figure 2-3a.*

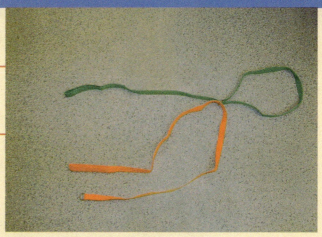

Figure 2-3a: The green leash is the one that will be around the dog's neck. The orange leash will be used to remove the green one. Note that the orange leash is looped through the ring on the green one.

Technical Action

2. You need to make a noose leash out of the orange leash.

Rationale/Amplification

2a. *See Figure 2-3b.*

Figure 2-3b: Making a loop with the orange leash.

Technical Action

3. Tighten the orange noose leash all the way down on to the ring on the green leash.

Rationale/Amplification

3a. This will enable you to remove the green leash by pulling on the orange leash. *See Figure 2-3c.*

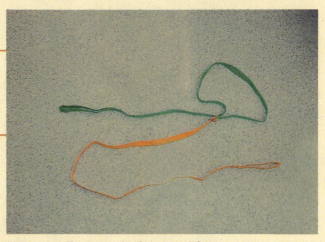

Figure 2-3c: The orange leash is now tightened down on the ring of the green leash.

Technical Action

4. Place the green noose leash around the dog's neck.

Rationale/Amplification

4a. Note that at this point you will have both leashes in your hand. *See Figure 2-3d.*

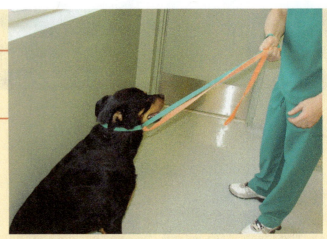

Figure 2-3d: The green leash is placed on the dog and the restrainer is holding both leashes in his hand.

Technical Action

5. To remove the green noose leash, release the green one while pulling on the orange one.

Rationale/Amplification

5a. *See Figure 2-3e.*

5b. Although not shown, this technique is used to remove a leash from an aggressive dog that you have placed in a cage or run.

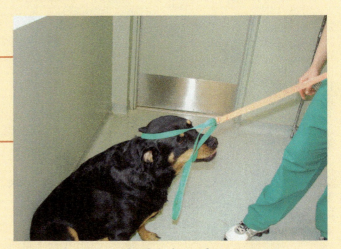

Figure 2-3e: The green leash is dropped while holding on to the orange one. This in a real situation could be used when an aggressive dog is led back in a cage or run and you cannot touch it to get the leash off. The orange leash is pulled to remove the green one.

Using a Noose Leash to Restrain a Dog Behind a Door

This technique is used when a fractious dog requires an injection in the rear limb and the dog is uncooperative. Three people are required for this procedure. You will need an assistant, plus someone to give the injection.

Note: This technique should not be performed in front of an owner.

PROCEDURE
Using a Noose Leash to Restrain a Dog Behind a Door

Technical Action

1. Place the noose leash around the dog's neck.

Rationale/Amplification

1a. You will need an assistant for this, plus someone to give the injection. In total, three people are required.

Technical Action

2. Use a run door.

Rationale/Amplification

2a. The door of an exam room or treatment room could also be used.

Technical Action

3. Lead the dog into the corner where the run door meets the wall and feed the leash through the hinge side of the door.

Rationale/Amplification

3a. *See Figure 2-4a.*

3b. If using an exam room door, the dog would be led to the hinge side of the door that will be opening into the room. The leash will then be fed through the hinge to an assistant. The assistant will pull the leash to get the dog as close to the door hinge as possible.

Figure 2-4a: Feeding the leash through the door hinge.

Technical Action

4. Pull on the leash to get the dog as close to the corner as possible.

Rationale/Amplification

4a. *See Figure 2-4b.*

Figure 2-4b: Pulling the leash to bring the dog up to the hinge.

Technical Action

5. An assistant will need to swing the run door tightly toward the wall, up against the dog.

Rationale/Amplification

5a. This will make a modified "squeeze cage." *See Figure 2-4c.*

5b. If using the exam room door, after you have handed the leash to your assistant, you can now push the door up close to the dog.

5c. This needs to be done quickly as the dog while squeezed behind the door will struggle, and choking the dog must be prevented.

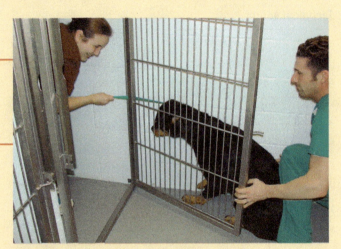

Figure 2-4c: Swinging the cage door up against the dog.

Technical Action

6. Another assistant will then come behind to give the injection in the hind limb.

Rabies Pole

The **rabies pole** is also known as a restraint pole or capture pole. Although municipal rabies control officers and shelter employees might have the occasion to use these often, veterinary hospitals use these only as a last resort. This device has a rigid pole with a noose attached to allow the restrainer distance and security away from the animal. *Note:* This should never be used in the presence of an owner.

The rabies pole is used to capture a dog or cat when you cannot put your hands on it. This could be used to remove a dog or cat from a cage or to capture an animal that got loose in the hospital. This may be the restraint necessary to give a sedative to allow for a physical examination or treatment to the animal.

SAFETY ALERT
One must be very careful not to choke an animal with this device.

PROCEDURE
Using the Rabies Pole

Technical Action

1. This technique is used to capture an aggressive loose dog or cat.

Rationale/Amplification

1a. In this case, the rabies pole is used instead of the noose leash. This procedure can also be used to remove a dog or cat from an enclosure or put one in an enclosure. *See Figure 2-5a.*

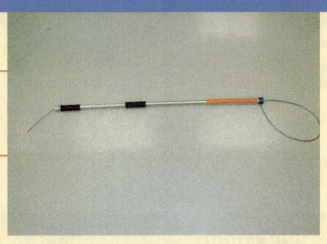

Figure 2-5a: Rabies pole.

Technical Action

2. Once the noose is around the neck, quickly tighten it.

Rationale/Amplification

2a. Anyone using this pole should familiarize themselves with the tightening mechanism of the noose before using it. *See Figure 2-5b.*

2b. Expect the animal to act violently to this maneuver. They will likely twist and flail.

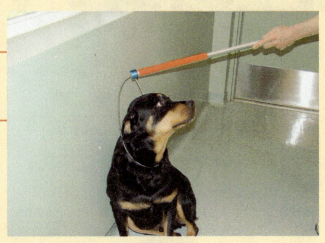

Figure 2-5b: Capturing the dog with the noose.

Technical Action

3. You should hold the pole at its maximum length firmly with both hands to give yourself distance from the dog or cat.

Rationale/Amplification

3a. *See Figure 2-5c.*

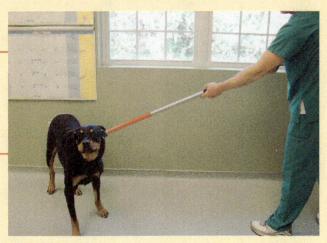

Figure 2-5c: Correctly holding the rabies pole.

Technical Action

4. You can now lead the animal to an enclosure or to a door to use it for restraint.

Blankets and Towels

 Blankets and towels are used to throw over an uncooperative animal in a cage or on an exam table. This technique can be used to provide a visual barrier so the animal cannot see what's coming next. Blankets and towels can also be used to gain control of a dog or cat to protect the hands from being bitten.

PROCEDURE
Using a Blanket or Towel

Technical Action

1. Obtain a large towel free of holes.

Rationale/Amplification

1a. Feet can get caught in the holes.

1b. If it is a thin towel, it should be folded over (doubled).

1c. A blanket of appropriate size can be used instead.

Technical Action

2. Open the cage door and throw the towel over the cat's (or dog's) body, including the head.

Rationale/Amplification

2a. This needs to be done in one swift movement. *See Figure 2-6a.*

Figure 2-6a: Approaching a cat in a cage with a towel.

2b. Covering the head prevents the animal from seeing what you are going to do next. *See Figure 2-6b.*

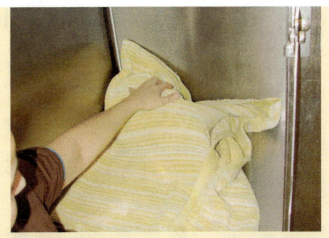

Figure 2-6b: Covering the cat with the towel.

Technical Action

3. Grasp the animal near the head with one hand and lift the body with the other.

Rationale/Amplification

3a. This must be done quickly. *See Figure 2-6c.*

Figure 2-6c: Grasping the cat using the towel to protect your hands.

Technical Action

4. Place the animal on the examination table.

Technical Action

5. From here, depending on what procedure will be performed and the cooperation of the animal, it can be restrained continuing to use the towel.

Rationale/Amplification

5a. Alternatively, once on the table, the restrainer can get a better hold without the towel.

Leather Gloves

Most veterinary hospitals have a pair of leather gloves; however, they are rarely used. These should be used only as a last resort as these alone can be frightening to an animal. Use of leather gloves:

- Allows a restrainer to put their hands on a very aggressive small dog or cat when all other restraint options have been exhausted.

 - Protects the restrainer's hands from being bitten.

- Allows a small dog or cat to be removed from a cage or from a crate of which the top has been removed, or a loose animal to be picked up off the floor.

PROCEDURE
Using Leather Gloves

Technical Action

1. Put the gloves on your hands. Put one on without pushing the fingers to the end of the glove. Don the other glove completely.

Rationale/Amplification

1a. They are "one size fits all." *See Figure 2-7a.*

1b. By allowing the end of one glove to dangle at the end of your fingers, you can use this as a mode of distraction with the animal. If the animal should try to bite, your fingers will not be bitten.

Figure 2-7a: Leather gloves. Note that the fingers of the glove are bent back to show that the glove is not completely donned. This hand can be used to distract the dog.

Technical Action

2. Grasp the animal around the neck with the completely gloved hand while distracting the animal with the other hand.

Rationale/Amplification

2a. All of this will need to be done quickly. *See Figure 2-7b.*

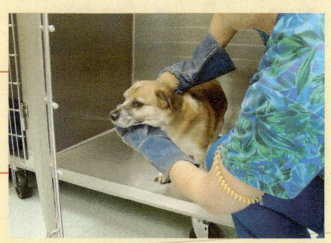

Figure 2-7b: The dog is now grasped with both hands.

Technical Action

3. Once the neck is grasped, the animal can be lifted using the other hand for support.

Technical Action

4. The animal can now be taken to wherever it needs to go.

Cat Bag

A **cat bag** is useful for restraint because it prevents the cat from scratching. The head is still exposed, however. These bags are typically made of nylon. The cat bag is used to:

- Restrain a cat so that personnel do not get scratched.
- Provide access to specific areas of the body through various zippered openings allowing for injections or venipuncture.
- To provide restraint while anesthetizing a cat with an anesthetic mask.

PROCEDURE
Using a Cat Bag

Technical Action

1. Obtain the cat bag you will be using and unzip the top all the way open.

Rationale/Amplification

1a. *See Figure 2-8a.*

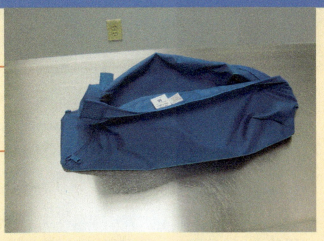

Figure 2-8a: Nylon cat bag.

Technical Action

2. Scruff the cat and lift it into the bag.

Rationale/Amplification

2a. This should be done in one swift motion.

Technical Action

3. Wrap the Velcro strap around the cat's neck and immediately zip up the bag.

Rationale/Amplification

3a. This too must be done swiftly before the cat realizes what is going on. *See Figures 2-8b and 2-8c.*

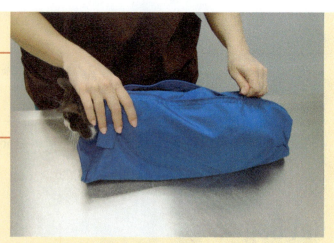

Figure 2-8b: Cat placed in the cat bag.

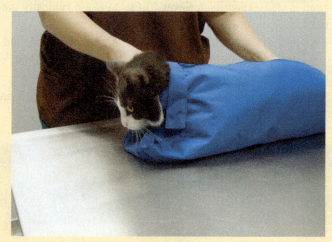

Figure 2-8c: The bag is zipped and the Velcro strap is secured around the cat's neck.

Technical Action

4. The various zippers on the bag can be used for different procedures.

Technical Action

5. When time to remove the cat from the bag, the Velcro strap should be removed first, then unzip the bag.

Rationale/Amplification

5a. There are two zippers on the front of the bag, which can be used to expose a forelimb for cephalic vein venipuncture. *See Figure 2-8d.*

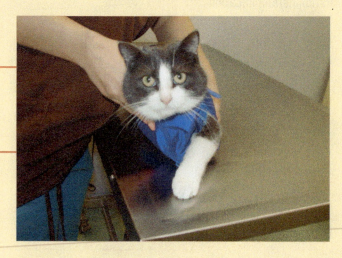

Figure 2-8d: Exposing a front limb using one of the zippered openings.

Dog Muzzles

A **muzzle** provides restraint to the head when common restraint methods are not enough. Interestingly, in some cases, placing a muzzle on a dog will create submission almost as if the dog has decided to give up once it figures out it can't bite. These can be purchased made of leather or nylon or can be fashioned from a gauze roll or a noose leash. The purpose for using a muzzle is to prevent a dog from biting while using a standing, sitting, or sternal restraint method when performing clinical procedures.

PROCEDURE
Applying a Commercial Dog Muzzle

Technical Action

1. The dog should be in a sitting or sternal position on an examination table or floor depending on the size of the dog.

Rationale/Amplification

1a. Sometimes you will need an assistant to restrain the dog while the muzzle is applied.

Technical Action

2. Come from behind the dog's head with the muzzle in one hand.

Rationale/Amplification

2a. The hand should grasp the side of the muzzle that does not have the strap. *See Figure 2-9a.*

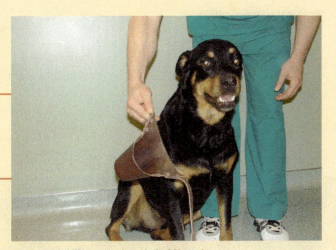

Figure 2-9a: The restrainer is holding the muzzle by the buckle end of the muzzle.

Technical Action

3. Bring the muzzle up to the dog's face and slip it on while grasping the strap with the other hand.

Rationale/Amplification

3a. *See Figure 2-9b.*

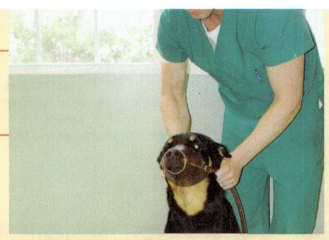

Figure 2-9b: The muzzle is placed on the dog's face and the strap grasped.

Technical Action

4. Secure the muzzle.

Rationale/Amplification

4a. If the muzzle is leather, it will have a buckle. If it is nylon it will be secured with Velcro, a buckle or a plastic clasp.

Technical Action

5. The proper fit would allow for one finger to be inserted under the strap.

Rationale/Amplification

5a. *See Figure 2-9c.*

5b. Make sure that the muzzle is not touching the dog's eyes.

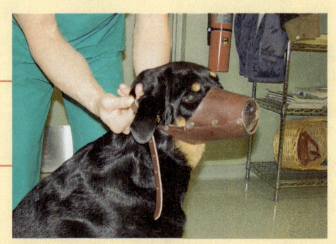

Figure 2-9c: The muzzle is checked to be sure it is not on too tight and that it is not touching the dog's eyes.

PROCEDURE
Applying a Noose Leash Muzzle

Technical Action

1. The dog should be in a sitting or sternal position on an examination table or floor depending on the size of the dog.

Rationale/Amplification

1a. The presence of an assistant to restrain the dog while the muzzle is applied is recommended.

Technical Action

2. Make a loop in the noose leash.

Rationale/Amplification

2a. The loop should be larger than the dog's muzzle but not so large that you cannot tighten it quickly.

Technical Action

3. Bring the loop up to the face and tighten quickly with the tie underneath.

Rationale/Amplification

3a. Make sure you keep your hands and arms away from the dog's mouth to prevent being bitten. *See Figures 2-10a and 2-10b.*

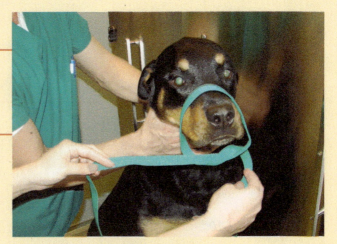

Figure 2-10a: Noose leash being used as a muzzle. Note that it is being tied under the dog's muzzle and the restrainer is steadying the head.

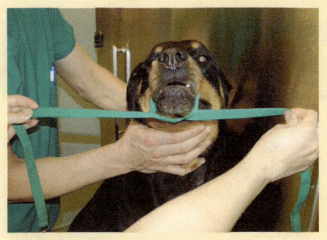

Figure 2-10b: The leash is being tightened.

Technical Action

4. Take the ends of the leash and pull them back behind the dog's head and tie.

Rationale/Amplification

4a. The straps will run under the ears. *See Figure 2-10c.*

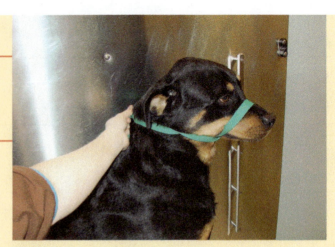

Figure 2-10c: The ends of the leash are gathered behind the dog's head below the ears.

Technical Action

5. The leash should be tied in a knot that will allow for quick release when needed. Tying the ends in a bow will work.

Rationale/Amplification

5a. *See Figure 2-10d.*

Figure 2-10d: The leash tied in a knot that can be easily grasped and quickly untied.

PROCEDURE
Applying a Gauze Dog Muzzle

Technical Action

1. The dog should be in a sitting or sternal position on an examination table or floor depending on the size of the dog.

Rationale/Amplification

1a. An assistant may be required with some dogs.

1b. This is a modification of the noose leash muzzle technique.

Technical Action

2. Obtain a roll of one inch or two inch gauze.

Rationale/Amplification

2a. The width that is chosen depends on the size of the dog's muzzle.

Technical Action

3. Make a loop in the gauze and approach the dog from behind.

Rationale/Amplification

3a. *See Figure 2-11a.*

Figure 2-11a: Making a loop large enough to fit over the dog's muzzle with rolled gauze. This is a large dog, so two inch gauze is used.

Technical Action

4. Place the loop on the dog's face with the tie on top.

Rationale/Amplification

4a. This needs to be done quickly before the dog realizes what is happening. *See Figure 2-11b.*

Figure 2-11b: The gauze is tightened on the top of the dog's muzzle.

Technical Action

5. Quickly tighten the loop, then cross the ends under the dog's face

Rationale/Amplification

5a. *See Figure 2-11c.*

Figure 2-11c: The gauze is then crossed under the dog's muzzle.

Technical Action

6. Bring the ends back behind the dog's head under the ears and tie.

Technical Action

7. The gauze should be tied in a knot that will allow for quick release when needed. Tyng the ends in a bow will work.

Rationale/Amplification

7a. *See Figure 2-11d.*

Figure 2-11d: The gauze is tied in a knot that can be easily grasped and quickly untied.

SAFETY ALERT ⚠

One must never assume that total safety is achieved with a muzzle in place. The muzzle can slip off. Also if a cat is not watched carefully, it will claw the muzzle off. Never forget that while the teeth can't be used as a weapon, the claws still can.

Cat Muzzles

Cat muzzles provide restraint to the head when common restraint methods are not enough. Interestingly, in some cases, as with the dog, placing a muzzle on a cat will create submission almost as if the cat has decided to give up once it figures out it can't bite. These types of muzzles are made of nylon. A muzzle is used on a cat to prevent the cat from biting while using a standing, sitting, or sternal restraint method when performing clinical procedures.

PROCEDURE
Applying a Cat Muzzle

Technical Action

1. The cat should be in a sitting or sternal position on an examination table.

Rationale/Amplification

1a. An assistant may be required when working with some cats.

Technical Action

2. Obtain a muzzle of appropriate size for the cat. Position the muzzle properly in your hands—the top of the muzzle is wider than the bottom.

Rationale/Amplification

2a. These come in small, medium, and large and will use Velcro to secure the muzzle to the head.

Technical Action

3. With the muzzle in both hands, approach the cat from behind.

Rationale/Amplification

3a. Have one strap in each hand. *See figure 2-12a.*

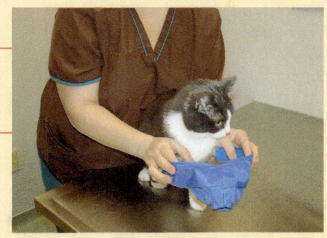

Figure 2-12a: Holding the cat muzzle with two hands, preparing to bring it up to the cat's face.

Technical Action

4. Bring the muzzle up to the cat's face in one swift motion.

Rationale/Amplification

4a. This needs to be done quickly so that the cat does not have time to react. *See Figure 2-12b.*

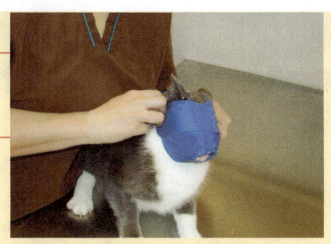

Figure 2-12b: Covering the cat's face with the muzzle.

Technical Action

5. Secure the muzzle.

Rationale/Amplification

5a. *See Figure 2-12c.*

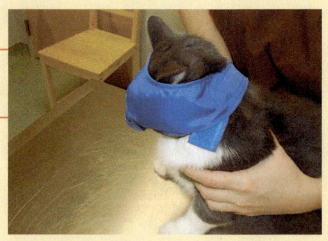

Figure 2-12c: Securing the Velcro straps.

SAFETY ALERT

One must never assume that total safety is achieved with a cat wrapped in a towel. The cat may be able to wiggle free using the front or rear paws. Also keep in mind that the head is not restrained so biting is possible.

Cat Burrito Restraint

The cat burrito restraint uses a towel in which the cat is wrapped tightly such that the feet are secured so the cat's claws cannot scratch the restrainer. This method is used to prevent a cat from scratching the restrainer while performing clinical procedures.

PROCEDURE
Applying Cat Burrito Restraint

Technical Action

1. Obtain a towel of appropriate size and place the cat in sternal recumbency on it.

Rationale/Amplification

1a. Larger cats require larger towels. The cat should be placed on one end of the towel. *See Figure 2-13a.*

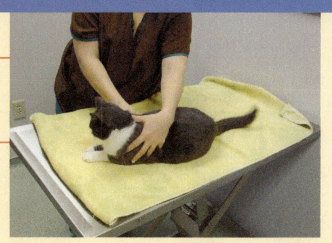

Figure 2-13a: Note that the cat is placed toward one end of the towel.

Technical Action

2. Fold the end of the towel that is near the tail over the cat's back and up to the neck.

Rationale/Amplification

2a. *See Figure 2-13b.*

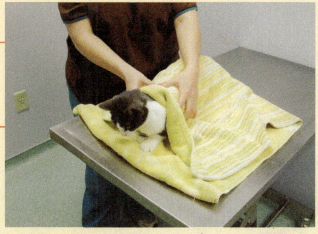

Figure 2-13b: The end of the towel is brought up over the cat's back.

Technical Action

3. Proceed to wrapping the cat by tucking the sides of the towel under the cat.

Technical Action

4. Bring both sides of the towel up over the back of the cat.

Rationale/Amplification

4a. If the cat's front paws are loose, causing a problem, depending on the procedure, the cat may have to be wrapped tighter to secure them. *See Figures 2-13c and 2-13d.*

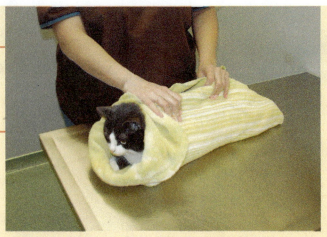

Figure 2-13c: The sides of the towel are gathered up on the cat's back.

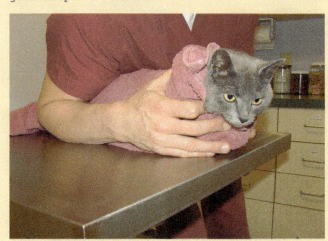

Figure 2-13d: Cat restrained with front feet contained.

Review Questions

1. Why are restraint devices used?
2. What is a noose leash?
3. Why should noose leashes be regularly cleaned?
4. Why is it best to have a fractious animal in a cage near the floor?
5. For what situation would restraining a dog behind a run door be useful?
6. What is another name for a rabies pole?
7. In what situation is a rabies pole used?
8. How is a towel used to remove an animal from a cage?
9. What situation would warrant the use of leather gloves?
10. Why is only one leather glove donned when using them for restraint?
11. What is a disadvantage of a cat restraint bag?
12. List four types of dog muzzles.
13. What precaution must be taken when a dog has a muzzle on?
14. How is a commercial cat muzzle different from a dog muzzle?
15. What precautions must be taken when using a cat muzzle?
16. What is the purpose of the cat burrito restraint technique?

Bibliography

Crow, S., & Walshaw, S. (1997). *Manual of clinical procedures in the dog, cat and rabbit*. Philadelphia: Lippencott-Raven.

McCurnin, D., & Bassert, J. (2006). *Clinical textbook for veterinary technicians* (6th ed.). St. Louis: Elsevier Saunders.

Sheldon, C., Sonsthagen, T., & Topel, J. (2008). *Animal restraint for veterinary professionals*. St. Louis: Mosby Elsevier.

4. So that way if they are agressive they ~~can~~ aren't close to your face.

10. They use the leather glove to grasp the animal around the neck and use the non-gloved hand to distract the animal.

11. A disadvantage to a cat bag would be that if you put them in wrong they could easily get their paws out and scratch you.

Chapter 3

Restraint of Dogs

If a dog will not come to you after having looked you in the face, you should go home and examine your conscience.

—Woodrow Wilson

Objectives

- Identify behavioral characteristics of the dog.
- Identify different behaviors in dogs that warrant different initial approaches and restraint techniques.
- Identify potential complications to a given restraint technique.
- Describe restraint techniques used on the dog to accomplish various procedures.

Key Terms

brachycephalic
cystocentesis
dyspnea
proptosis

Restraint of the Dog

Most dogs are very accustomed to being handled. In a veterinary clinic, restraint is required to accomplish procedures such as a physical examination, venipuncture, or a nail trim. Proper restraint allows a procedure to be performed without injury to the dog or the humans performing the procedure.

Guidelines for Restraint of the Dog

The dog can be easy to work with and most are fairly cooperative participants in procedures due to their desire for human companionship and experience with handling. However, the guidelines below should be followed:

- Always introduce yourself to the dog first before doing anything. Speak to the dog in a friendly, calm voice. Use the dog's name when you talk to it. When you approach the dog, present one hand with the palm down. Allow the dog to come to you rather than advancing on the dog. If the dog is friendly, it will come to you and you should see a wagging tail. It may even lick your hand. This is a positive sign that the dog will likely cooperate with basic restraint.
- It is helpful when handling a dog to determine if it has had any obedience training by asking it to sit or stand. If it responds to commands this can make restraint easier. If for example it responds to the command to sit, then forcing it to sit will not be necessary.
- Realize that the type of restraint may be dictated by how comfortable with a certain technique the animal is. Some dogs are very stoic and will not react to a procedure that may be uncomfortable. Less forceful restraint might be what is required. On the other hand, some dogs become very excited just seeing a pair of nail clippers. These dogs may need more than one restrainer. In both cases, restraint such that the other person doesn't get hurt is paramount.
- The goal should always be that the restraint and the procedure can be performed in such a way that it is atraumatic and non-stressful to everyone including the dog.
- Some dogs cooperate much better out of sight of the owner. In those cases, one may ask (or tell) the owner that the dog will be taken back to the treatment area for venipuncture for a heartworm test, for example, and returned to the exam room when finished.
- Rewarding the dog with a treat after the procedure ends the experience on a positive note.

It is the personnel's responsibility to minimize injury to the dog. Unfortunately, there are undesirable complications to the pet associated with restraint, which veterinary technicians and assistants need to be aware of and avoid at all costs. These include:

- Trauma to the patient such as sprains and muscle soreness due to excessive force
- Hyperthermia from excessive struggle and excitement
- **Dyspnea** (labored breathing) especially in **brachycephalic** (short, broad head) breeds
- **Proptosis** (displacement) of the eye due to excessive force around the neck of dogs with protruding eyes (i.e., pugs)
- Emotional stress

The health status of the dog may predispose it to pain and injury during restraint. Obviously a dog with a previous back injury will require gentle handling. Arthritic dogs necessitate special consideration since many areas of their body may be painful and restraint techniques could elicit aggression. Dyspneic dogs will require careful restraint so as not to put them in respiratory distress.

Injury to restrainers is possible as well. Scratches from struggling or kicking can occur during handling. Bites are also an inherent danger. Lifting a heavy dog improperly can cause back injury. Personnel should never try to impress others by lifting an animal that is too heavy. Large dogs should be lifted using two people, one for the front end and the other for the back end. Do not be too proud to ask for help.

There may be situations in which the most experienced restrainers cannot get a patient to cooperate for a procedure using proper restraint techniques. In these cases the chance of injury to the patient or personnel increases. In these cases it is in the best interest of the patient and personnel to use chemical restraint.

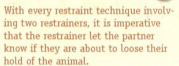

SAFETY ALERT

With every restraint technique involving two restrainers, it is imperative that the restrainer let the partner know if they are about to loose their hold of the animal.

Restraint of the Dog in the Standing Position

Purpose

Dogs may need to be restrained in a standing position to perform procedures such as a physical examination, injections, venipuncture, ophthalmic examinations, and so on. This can be performed with the dog on the examination table or on the floor. The location chosen is determined by the person performing the procedure and the animal itself. For example, if the dog is extremely large, having it stand on the table may make it "too tall" to reach to perform proper restraint or for the other person to perform the procedure at hand. Alternatively, some dogs may be uncooperative on an examination table but become totally cooperative on the floor.

Complications

- Patient becomes dyspneic from too strong a hold around the neck
- One may have trouble controlling the limbs of the dog
- Proptosis of an eye
- Injury to personnel

Equipment

- Noose leash
- Examination table depending on the dog's size

PROCEDURE
Restraint of the Dog in the Standing Position

Technical Action

1. Place a noose leash on dog.

Rationale/Amplification

1a. This gives you control over the dog if you need it.

Technical Action

2. Wrap one arm around the dog's head close to the jaw and the other arm under the abdomen in the space between the **caudal** aspect of the ribs and the hind limbs.

Rationale/Amplification

2a. The dog's face should always be kept away from the person performing the procedure to reduce the chance of the other person getting bitten. *See Figure 3-1a.*

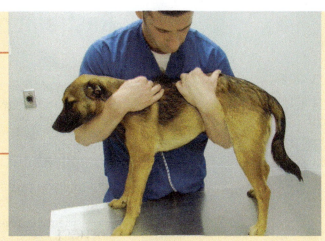

Figure 3-1a: Restraining a dog in the standing position.

Technical Action

3. Hold the dog close to your body.

Rationale/Amplification

3a. How strong the hold is depends on the cooperation of the dog and the unpleasantness of the procedure.

3b. If the dog is a small one, the dog may cooperate better if its feet are not touching the exam table.

3c. If a rectal examination is being performed you may be asked to hold up the tail. This will require you to remove your arm from the abdomen. *See Figure 3-1b.*

3d. If someone is performing a physical examination, you may be asked to move your arms to allow access to those parts of the body. *See Figure 3-1c.*

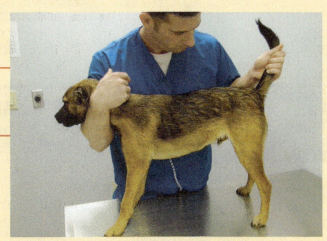

Figure 3-1b: Standing restraint holding the tail up.

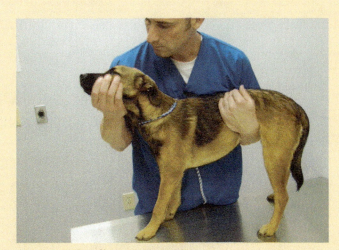

Figure 3-1c: Standing restraint with arms away from the chest for auscultation.

Restraint of the Dog in the Sitting Position

Purpose

To provide restraint for the same procedures listed under standing restraint. As with the standing restraint, a large dog may sit "too tall" to be restrained on an examination table.

Complications

- Injury to personnel
- Proptosis of an eye
- Patient becomes dyspneic from too strong a hold around the neck

Equipment

- Noose leash
- Examination table depending on the size of the dog

PROCEDURE
Restraint of the Dog in the Sitting Position

Technical Action

1. Put a noose leash on the dog.

Rationale/Amplification

1a. This gives you control over the dog if you need it.

Technical Action

2. If the dog is standing, ask the dog to sit. If it doesn't, gently press down on the hindquarters to encourage it to sit.

Rationale/Amplification

2a. Be careful when you try to get a dog with arthritic hips to sit. It may be painful and elicit an aggressive response.

Technical Action

3. Place one arm around the dog's neck and the other arm around the dog's chest.

Rationale/Amplification

3a. The dog's face should always be kept away from the person performing the procedure to reduce the chance of the other person getting bitten. *See Figure 3-2.*

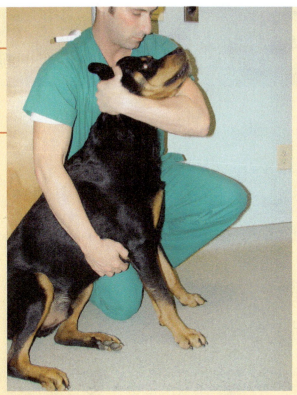

Figure 3-2: This is a large dog so the dog is being restrained in the sitting position on the floor.

Restraint of the Dog in Sternal Recumbency

Purpose

To perform procedures such as a physical examination, injections, venipuncture, and ophthalmic examinations. In some cases, the dog may be more cooperative if it is restrained in this position.

Complications
- Injury to personnel
- Proptosis of an eye
- Patient becomes dyspneic from too strong a hold around the neck

Equipment
- Noose leash
- Examination table depending on the dog's size

PROCEDURE
Restraint of the Dog in Sternal Recumbency

Technical Action

1. Put a noose leash on the dog.

Rationale/Amplification

1a. This gives you control over the dog if you need it.

Technical Action

2. Begin with the dog sitting.

Technical Action

3. Place one arm around the dog's neck and place the other arm around the dog's back to grasp the forelimbs.

Rationale/Amplification

3a. The dog's face should always be kept away from the person performing the procedure to reduce the chance of the other person getting bitten. *See Figure 3-3a.*

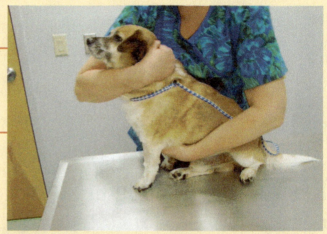

Figure 3-3a: Restraint of a small dog in the sitting position on an examination table.

Technical Action

4. Push down on the dog's back with your body. This will encourage the dog to lie down.

Rationale/Amplification

4a. Apply gentle pressure so as not to injure the dog's back. You may need to pull the forelimbs forward to get the dog to lie down. *See Figure 3-3b.*

4b. Two people may be required to restrain a large dog. One person would restrain the front end of the dog as previously described, while a second person uses their hands to keep the hindquarters on the table or floor.

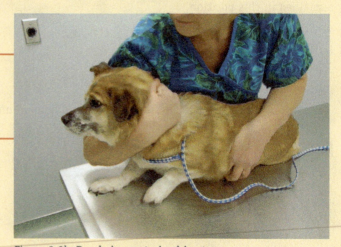

Figure 3-3b: Dog being restrained in sternal recumbency. Note that the restrainer is leaning over the dog with her left arm cradling it close to her body.

Technical Action

5. From this position the head can be examined.

Rationale/Amplification

5a. See section on head restraint.

Restraint of a Dog's Head for Procedures Performed by Someone Else

Purpose

To restrain a dog's head for someone who is to perform an ophthalmic, otic, or oral examination, or to administer otic, ophthalmic, or oral medications.

Complications

- Dog may panic if it cannot pant
- Bite injuries to personnel

Equipment

- Noose leash
- Examination table depending on dog's size

PROCEDURE
Restraint of a Dog's Head for Procedures Performed by Someone Else

Technical Action

1. Put a noose leash on the dog

Rationale/Amplification

1a. This gives you control over the dog if you need it.

Technical Action

2. Place the dog in a sitting, standing, or sternal recumbent position.

Rationale/Amplification

2a. The position chosen depends on the size of the dog and its temperament.

Technical Action

---

3. The dog's head is held with both hands by placing the palms of the hands on the caudal portion of the mandible, with fingers holding the face.

Rationale/Amplification

3a. *See Figure 3-4a.*

3b. Caution must be taken to not get your fingers in the way of the mouth.

3c. While the dog can be in any position to hold the head with both hands, the sternal position may be best with some dogs as the forearms of the restrainer can be placed on either side of the dog's chest for added control. *See Figure 3-4b.*

3d. If the dog is small, one hand can be used. The hand holds the muzzle with the thumb under the jaw and the fingers placed over the top of the muzzle, essentially holding the muzzle like a hamburger!

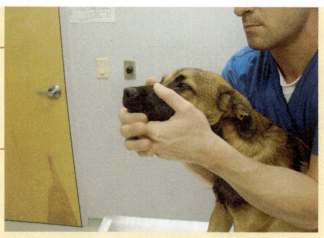

Figure 3-4a: Holding a dog's head with two hands.

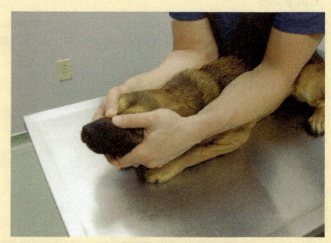

Figure 3-4b: While having the dog in sternal recumbency while holding the head, the restrainer can use his forearms to control the dog.

Single Person Restraint of a Dog's Head When Performing a Procedure to the Head

②Purpose

To restrain a dog's head for procedures that will be performed to the head such as applying otic, ophthalmic, or oral medication and you have no assistants to help you.

Complications

- Less control over the patient
- Bite injuries to restrainer

Equipment

- Noose leash
- Examination table depending on the size of the dog

PROCEDURE
Single Person Restraint of a Dog's Head When Performing a Procedure to the Head

Technical Action

1. Place a noose leash on the dog.

Rationale/Amplification

1a. This will give you control over the dog if you need it.

Technical Action

2. Place the dog in a sitting position or in sternal recumbency.

Rationale/Amplification

2a. For any dog less than 100 percent cooperative, the sternal position is easiest.

Technical Action

3. Hold the dog's head with one hand holding the muzzle from underneath. The other hand holding the medication will be placed over the dog's head in the case of eye medication.

Rationale/Amplification

3a. *See Figure 3-5a.*

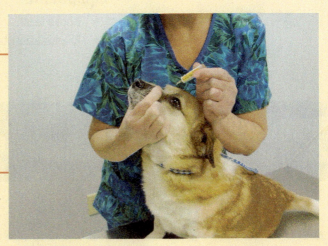

Figure 3-5a: Applying ophthalmic ointment to a dog's eye without assistance.

Technical Action

4. For oral medication, one arm will be wrapped around the dog's neck near the mandible. The other hand will reach around the head to apply the medication.

Rationale/Amplification

4a. *See Figure 3-5b.*

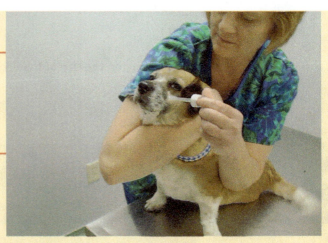

Figure 3-5b: Administering oral medication to a dog without assistance.

Restraint of a Dog in Lateral Recumbency

Purpose

To restrain a dog in a lateral recumbency position for procedures such as urinary **catheterization**, injections, **cystocentesis** (puncture of the bladder to obtain a urine sample), and venipuncture of the lateral saphenous vein.

Complications

- Injury to personnel
- Dyspnea from pressure on the dog's neck
- Back or neck injury if dog is allowed to struggle

Equipment

- Noose leash
- Examination table depending on the size of the dog

PROCEDURE
Restraint of a Dog in Lateral Recumbency

Technical Action

1. Put a noose leash on the dog.

Rationale/Amplification

1a. This will give you control over the dog if you need it.

Technical Action

2. Place the dog in the standing position.

Rationale/Amplification

2a. The dog could also be placed in lateral recumbency from a sternal position.

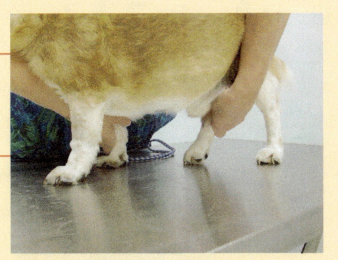

Technical Action

3. Place your right arm across the dog's neck and reach between the front legs to grasp the dog's right forelimb in your right hand.

Rationale/Amplification

3a. This will place the dog in right lateral recumbency. *See Figure 3-6a.*

3b. If the dog is in sternal recumbency, the right front limb will not be exposed for easy access so grasping the right forelimb will be more difficult.

3c. Alternatively you can place your right arm over the dog's neck and reach under the chest to place the dog's right forelimb in your right hand.

Figure 3-6a: This photograph is showing where the restrainer's hands need to be in order to put this dog in right lateral recumbency.

Technical Action

4. Place your left arm over the dog's back, reaching for the dog's right rear limb. Grasp the limb just proximal to the hock.

Rationale/Amplification

4a. If the dog is in sternal recumbency, the right rear limb will be tucked under the dog and will be more difficult to grasp.

Technical Action

5. With the dog's body close to yours, gently lift the limbs while allowing the dog's body to move to the table. It should now be on its right side.

Rationale/Amplification

5a. This needs to be done quickly and in one motion as some dogs will panic when they feel they are losing their footing. Placing the dog on the table must be done gently. *See Figure 3-6b.*

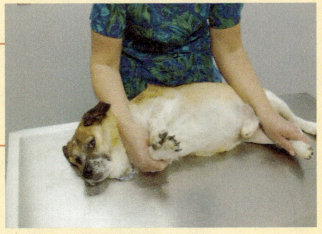

Figure 3-6b: This is the way the dog is restrained to maintain lateral recumbency.

Technical Action

6. Allow the dog to relax for a couple of seconds, not releasing the limbs from your grasp. Your right arm can now be used to place pressure on the dog's neck if needed for more control.

Technical Action

7. If the dog is large, two people may be required for this restraint. One person would be in control of the front limbs, while the other would control the hind limbs.

Rationale/Amplification

7a. *See Figures 3-6c and 3-6d.*

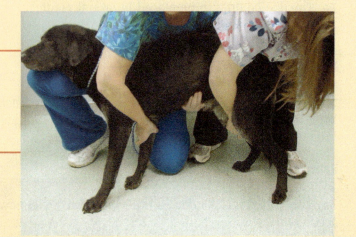

Figure 3-6c: This photograph shows two people preparing to put a large dog in lateral recumbency.

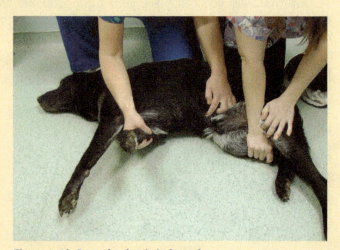

Figure 3-6d: Once the dog is in lateral recumbency, one person controls the front limbs, while the other person controls the hind limbs.

Technical Action

8. Use caution when releasing the dog from lateral recumbency. The hind legs should be released first followed by the front.

Restraint of the Dog for Cephalic Venipuncture

Purpose

- This technique provides restraint for venipuncture of the forelimb

Complications

- Injury to limb
- Bite injuries to personnel

Equipment

- Noose leash

PROCEDURE
Restraint of the Dog for Cephalic Venipuncture

Technical Action

1. Place a noose leash on the dog.

Rationale/Amplification

1a. This gives you control of the dog if you need it.

Technical Action

2. Restrain the dog in sternal recumbency.

Rationale/Amplification

2a. Some dogs are more comfortable in the sitting position for this procedure. Positioning the limb and occluding the vessel is the same regardless of the dog's position.

Technical Action

3. For left cephalic venipuncture, place yourself on the dog's right side. Wrap your right arm around the dog's neck.

Rationale/Amplification

3a. Make sure the dog's face is directed away from the person performing the venipuncture.

Technical Action

4. Hold the dog's left forelimb with its elbow in the palm of your hand and extend the limb forward toward the person performing the procedure.

Rationale/Amplification

4a. At this point, the other person should grasp the leg to prepare the site for the venipuncture.

Technical Action

5. With the elbow in your palm, rotate your thumb up so it is on top of the limb at the bend of the elbow (where the radius meets the humerus). Pushing down with the thumb, rotate it to the left.

Rationale/Amplification

5a. The thumb helps to stabilize the vessel and occlude the vessel so it can be visualized. *See Figure 3-7.*

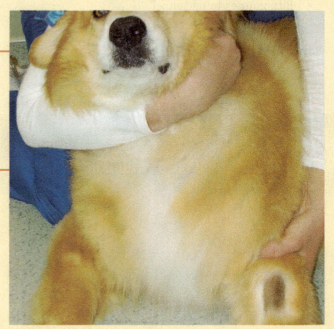

Figure 3-7: The area where the cephalic vein is located has been shaved in this dog to make it more visible. Shaving isn't routinely done for cephalic venipuncture. Note where the restainer's thumb is.

Technical Action

6. The venipuncture can now begin.

Rationale/Amplification

6a. The person performing the venipuncture will tell you when they want you to release your thumb prior to withdrawing the needle.

Restraint for Venipuncture of the Lateral Saphenous Vein

Purpose

• For venipuncture of the lateral saphenous vein

Complications

- Bite injuries to personnel
- Injury to the patient

Equipment

- Noose leash
- Examination table depending on the size of the dog

PROCEDURE
Restraint for Venipuncture of the Lateral Saphenous Vein

Technical Action

1. Place a noose leash on the dog.

Rationale/Amplification

1a. This gives you control of the dog if you need it.

Technical Action

2. Place the dog in lateral recumbency.

Rationale/Amplification

2a. This technique can also be performed with the dog standing. Right lateral recumbency will be described.

Technical Action

3. Restrain the dog in lateral recumbency as described previously. Rather than using your left hand to stretch and hold the rear limbs, your hand will be used to occlude the vein.

Rationale/Amplification

3a. In a cooperative dog, the fact that both hind limbs are not held is typically not a problem.

Technical Action

4. Your left hand will hold the limb tightly in the area just distal to the stifle, which will occlude the vein.

Rationale/Amplification

4a. The vein runs on the lateral surface of the hind limb and is visualized in the area between the distal stifle and proximal hock.

4b. *See Figure 3-8a.*

4c. The vein is now visualized. *See Figure 3-8b.*

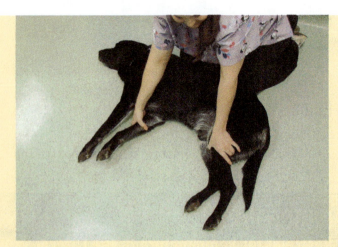

Figure 3-8a: This photograph shows restraint of a large dog for lateral saphenous vein venipuncture.

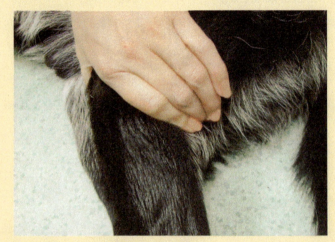

Figure 3-8b: Close up of the lateral saphenous vein. It can be seen crossing diagonally from left to right. Note where the restrainer's hand is.

Technical Action

5. The venipuncture can now begin.

Rationale/Amplification

5a. The person performing the venipuncture will tell you when they want you to release your hold on the limb prior to withdrawing the needle.

Restraint for Jugular Venipuncture

Purpose

- Provide restraint to obtain a large quantity of blood quickly

Complications

- Bite injury to personnel
- Injury to patient's neck from struggling

Equipment

- Noose leash

PROCEDURE
Restraint of the Dog for Jugular Venipuncture

Technical Action

1. Place a noose leash on the dog.

Rationale/Amplification

1a. This gives you control of the dog if you need it. The noose leash will need to be moved up toward the dog's jaw so it is out of the way during the procedure or it may need to be removed.

Technical Action

2. Place the dog in a sitting position.

Rationale/Amplification

2a. This restraint can also be done with the dog in sternal recumbency or standing position, depending on the size of the dog. With large dogs, sometimes this restraint is best done on the floor with the dog in the sitting position.

Technical Action

3. Hold the head up, away from the chest. This can be achieved by cupping your hand underneath the muzzle and by pushing the head up toward the ceiling.

Rationale/Amplification

3a. The person doing the procedure will direct the exact position of the head for maximum visualization of the vein. The vein in this picture is outlined in black. *See Figure 3-9.*

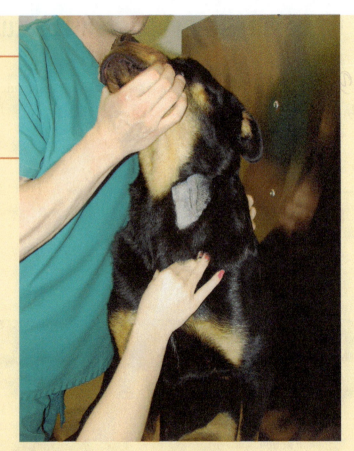

Figure 3-9: In this photograph, the jugular vein is between the black lines. The neck has been shaved for the purpose of showing this vein. It is not routine for the area to be shaved for venipuncture.

Technical Action

4. Your other arm should be wrapped over the dog's back.

Technical Action

5. For a small dog, it may be necessary to hold the dog right at the edge of the table, holding the head up as described above and using the other hand to grasp both front legs and extend them down off the edge of the table.

Rationale/Amplification

5a. Pulling the legs down and tilting the head up provides visualization of the vein. This technique does not provide the best restraint of the dog's body.

Review Questions

1. Give examples of procedures that would require restraint of a dog in a veterinary clinic.
2. Discuss situations that dictate different restraint techniques.
3. Describe the correct initial approach to a dog that you do not know.
4. List undesirable complications associated with dog restraint.
5. List injuries that personnel can sustain during restraint.
6. For restraining a dog in a standing position, what dictates whether the dog should be on the table or on the floor?
7. For what reasons may a dog be restrained in sternal recumbency?
8. What venipuncture is typically performed with a dog in lateral recumbency?
9. How is the lateral saphenous vein occluded by the restrainer?
10. Why would the jugular vein be chosen over the cephalic vein for venipuncture?

look @ comps

Bibliography

Crow, S., & Walshaw, S. (1997). *Manual of clinical procedures in the dog, cat and rabbit*. Philadelphia: Lippencott-Raven.

McCurnin, D., & Bassert, J. (2006). *Clinical textbook for veterinary technicians* (6th ed.). St. Louis: Elsevier Saunders.

Sheldon, C., Sonsthagen, T., & Topel, J. (2007). *Animal restraint for veterinary professionals*. St. Louis: Mosby Elsevier.

5. bites and scratches. (other injuries depending on the animal size and how agressive it is.)

6. The weight and size of the dog.

10. You get a faster blood flow with the jugular vein.

2. It depends on what procedure the animal gets that day that'll determine the restraint technique.

Chapter 4

Restraint of Cats

A cat sees no good reason why it should obey another animal, even if it does stand on two legs.

—Sarah Thompson

Objectives

- Identify behavioral characteristics unique to the cat.
- Identify different behaviors in cats that warrant different initial approaches and restraint techniques.
- Identify potential complications to the restraint technique.
- Describe restraint techniques used on the cat to accomplish various procedures.

Key Term

scruffing

Restraint of the Cat

Someone once said that cats are not small dogs. This is an understatement when it comes to restraining cats. Although cats are accustomed to handling just as the dog is, the difficulty lies in the fact that cats are not typically accustomed to traveling and to being in strange places. Rarely are cats taken for car rides to the park on a sunny day. By the time the cat gets to the clinic, it is already upset by the car ride getting there. Then, it is brought into a strange place. To add insult to injury, the cat is typically facing other cats, dogs, and strange smells the moment it reaches the reception area. Understanding this experience from the cat's vantage point will go a long way in providing the empathy required to successfully perform procedures on it.

Guidelines for Restraint of the Cat

Some say there is an art to handling cats. Those who are successful in handling cats have a basic understanding of cats' behavior. Although cats have a social structure when in groups, they are not social to humans in the same ways dogs are. The following guidelines should make interactions easier and safer:

- Minimal restraint should always be used unless the cat's behavior forces you to do otherwise. Do not automatically assume the worst with a cat. There are some cats that require virtually no restraint for a physical examination.

- Always introduce yourself to the cat before performing any procedure. This is accomplished by using a low voice, talking to the cat by its name. Extend your hand palm down, and allow the cat to sniff it. Taking an aggressive stance with the cat right off the bat can start the procedure off on the wrong foot and set a negative tone for the rest of the procedure and the cat's stay at the clinic.

- Cats typically give subtle warnings that they are getting upset before they attack. Restrainers should be observant of these. For example, if a cat is beginning to get agitated, their pupils may become more dilated or their ears may rotate rearward. As their agitation increases, they will flick the end of their tail. As agitation increases, the tail flicking can become more forceful to the point where it can be heard tapping on the examination table. Some cats will emit a low-pitched growl. These are all warnings that the cat has had enough. At this point, one needs to hurry and finish the procedure or apply more restraint.

- All means of escape need to be considered when a cat is handled. Doors, closets, and cabinets need to be closed!

- Sometimes just **scruffing** a cat will be all the restraint that is required. This involves grasping the skin around the base of the neck just as mother cats do to carry and reprimand their kittens. The amount of force applied with the scruff depends on the attitude of the cat. Some cats require quite a strong hold during a procedure, while others will cooperate for procedures with a light scruff.

- If a cat is scruffed and begins to become agitated, a distraction technique may help. Tapping the cat on the top of the head may put its mind on something else. If that doesn't help, sometimes simply placing a towel over the cat's head will quiet the cat down.

- If a cat is brought into the examination room in a crate, the best thing to do is allow it to walk out on its own. If the cat is friendly but does not want to come out, you can grab it by the scruff and pull it out. This cannot be done with an agitated cat. The crate may need to be disassembled first, removing the top.

- For cats that are particularly fearful, dimming the lights in the examination room may help.

- Make sure that everything you need for the procedure about to be performed is available. You may not have much time if the cat becomes agitated.
- Cats can inflict serious wounds with bite wounds to hands being the most common. For this reason, extreme caution should be taken with all cats but particularly with agitated or feral cats. Bite wounds to hands can end a career in veterinary medicine.
- There will be times when chemical restraint will be the only option for restraining an aggressive cat. Pride should not get in the way of its use. This approach will be the kindest for the patient and the safest for personnel.

<table>
<tr><td>

⚠ SAFETY ALERT

With every restraint technique used, it is imperative that the restrainer let the other person know if they are about to loose their hold on the animal.

</td></tr>
</table>

Restraint of the Cat Using the One-Hand Hold

Purpose

To allow the administration of an injection when you have no one to assist you. This has also been called the pretzel hold.

Complications

- Injury to cat from excessive force
- Personnel may be scratched or bitten

Equipment

- Cage

PROCEDURE
Restraint of the Cat Using the One-Hand Hold

Technical Action

1. This procedure is best accomplished by first grasping the animal from an examination table.

Rationale/Amplification

1a. Have the cat's cage open so that the cat will go into it immediately after the injection.

1b. Make sure that you have the injection ready.

Technical Action

2. Scruff the cat with one hand and use your other hand to place one rear paw in the scuffing hand.

Rationale/Amplification

2a. Before adding the rear paw to the scruffing hand, you must be sure you have an adequate hold.

Technical Action

3. Situate the rear limb so that the hock is in your grasp.

Rationale/Amplification

3a. *See Figure 4-1.*

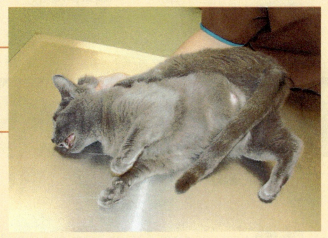

Figure 4-1: A one hand cat hold. Note that the cat's left hind limb is held in the restrainer's hand along with the cat's scruff.

Technical Action

4. Pick the cat up off the table with one hand and walk to the cage that you have opened. Have the cat facing the back of the cage. Pick up the syringe with your free hand.

Rationale/Amplification

4a. By giving the injection at the opened cage, if the cat reacts negatively it can be released directly into the cage.

Technical Action

5. Give the injection to the rear limb muscles that are presented when the limb is held with your scruffing hand.

Rationale/Amplification

5a. Release the cat into the cage. Be prepared to close the cage door quickly as some cats will immediately turn around and try to swat or bite you in retaliation!

Restraint of the Cat in Sternal Recumbency

Purpose

To allow for procedures such as venipuncture, giving injections, and physical examination.

Complications

- Injury to the cat due to excessive force
- Personnel may be scratched or bitten

Equipment

- Examination table

PROCEDURE
Restraint of a Cat in Sternal Recumbency

Technical Action

1. Scruff the cat and place it on an examination table.

Rationale/Amplification

1a. Some cats will automatically get in sternal recumbency when placed on a table.

Technical Action

2. While scruffing the cat, gently push downward to encourage the cat to sit.

Rationale/Amplification

2a. To avoid injury, do not place too much force on the cat's back.

Technical Action

3. Once the cat is sitting, gently push down on the cat's shoulders and front legs with your free hand.

Rationale/Amplification

3a. This should encourage the cat to extend the forelimbs. If that does not happen, you may need to grasp the front limbs and pull them forward.

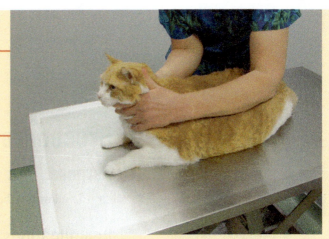

Figure 4-2: The cat restrained in sternal recumbency.

Technical Action

4. Once the cat is sternal you may need to place your arm around the cat to prevent it from getting up.

Rationale/Amplification

4a. *See Figure 4-2.*

4b. You may need to cradle the cat close to your body to maintain this position.

Restraint of the Cat in the Standing Position

Purpose

To allow for procedures such as a physical examination, nail trimming, or taking a temperature.

Complications

- Injury to the cat from excessive force
- Personnel may be scratched or bitten

Equipment

- Examination table

PROCEDURE
Restraint of the Cat in the Standing Position

Technical Action

1. Place the cat on the table and restrain using one hand around the cat's neck, while using the other hand to hold the cat under its abdomen.

Rationale/Amplification

1a. *See Figure 4-3.*

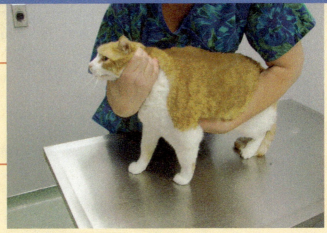

Figure 4-3: The cat restrained while standing.

Technical Action

2. Cradle the cat's body close to yours.

Procedure for Restraint of a Cat's Head

Purpose

- To allow for the ears, eyes, and mouth to be examined

Complications

- The cat becomes dyspneic from force around the neck
- Personnel can be bitten

Equipment

- Examination table

PROCEDURE
Restraint of a Cat's Head

Technical Action

1. Place cat in a sitting or sternal position on an examination table.

Technical Action

2. Grasp the head with your hands cupping your fingers under the jaw and your thumbs on top of the cat's head.

Rationale/Amplification

2a. *See Figure 4-4.*

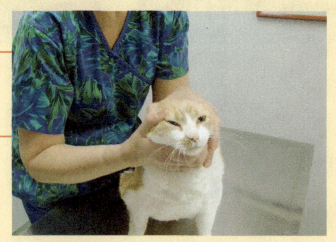

Figure 4-4: A cat's head being restrained for an examination.

"Cat Stretch" Restraint (Restraint of the Cat in Lateral Recumbency)

Purpose

To allow for a physical examination, giving injections, or other procedures that require a more secure method of restraint. This technique should not be performed in front of the owner.

Complications

- Injury to personnel from bites and scratches
- Injury to the cat due to excessive force

Equipment

- Examination table

PROCEDURE
"Cat Stretch" (Restraint of the Cat in Lateral Recumbency)

Technical Action

1. Place the cat on an examination table.

Technical Action

2. Scruff the cat with one hand and lift it off of the table enough to grasp both hind legs with your other hand.

Rationale/Amplification

2a. *See Figure 4-5a.*

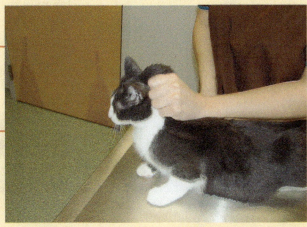

Figure 4-5a: This photograph shows a cat being scruffed.

Technical Action

3. Lay the cat on its side with the hind legs stretched rearward.

Rationale/Amplification

3a. Note that you will have no control over the front legs. *See Figure 4-5b.*

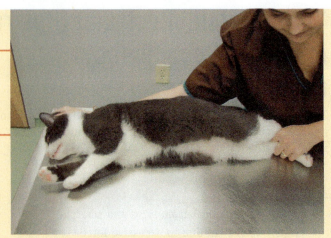

Figure 4-5b: A cat being restrained using the "cat stretch" technique. Note that the front limbs are not restrained.

Restraint of the Cat for Medial Saphenous Vein Venipuncture

Purpose

To obtain a blood sample from the medial saphenous vein. This technique should not be performed in front of the owner.

    Note: This technique can also be used to obtain a blood sample from a small uncooperative dog.

Complications

- Injury to personnel due to bites and scratches
- Injury to the cat due to excessive force

Equipment

- Examination table

PROCEDURE
Restraint of the Cat for Medial Saphenous Vein Venipuncture

Technical Action

1. Place the cat on an examination table.

Rationale/Amplification

1a. This is a modification of the cat stretch restraint technique.

Technical Action

2. Place the cat in the cat stretch restraint with the cat on its right side.

Rationale/Amplification

2a. This description will have the cat in right lateral recumbency.

Technical Action

3. Instead of holding both rear limbs with your left hand as in the stretch restraint, your left hand will be used to occlude the right medial saphenous vein.

Rationale/Amplification

3a. Note that you will have no control over the front limbs.

Technical Action

4. The person performing the venipuncture will grasp the right rear paw to extend it, as you will use the outer side of your left hand to press down on the medial surface of the thigh where it joins the torso.

Rationale/Amplification

4a. By pressing down in this area, the vein is occluded.

4b. The hand pressing down to occlude the vein is also holding the left rear leg and the tail out of the way. *See Figure 4-6.*

4c. You will be told when to release the vein.

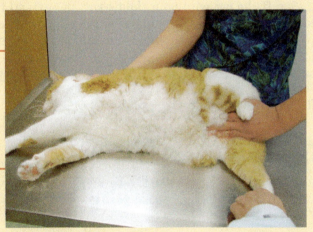

Figure 4-6: Cat being restrained for venipuncture of the medial saphenous vein. Note the position of the restrainer's hand on the right rear limb, which is occluding the vein.

Restraint of the Cat for Cephalic Venipuncture

Purpose
- To obtain a blood sample from the cephalic vein

Complications
- Injury to the cat due to excessive force
- Injury to personnel due to bites and scratches

Equipment
- Examination table

PROCEDURE
Restraint of the Cat for Cephalic Venipuncture

Technical Action

1. Place the cat in sternal recumbency on an examination table.

Rationale/Amplification

1a. This description will be for the left cephalic vein.

Technical Action

2. Scruff the cat with your right hand and extend the left front limb forward by grasping the elbow in the palm of your hand with your thumb on top of the elbow joint. The procedure for occlusion of the vein is the same as in the dog.

Rationale/Amplification

2a. Having the cat near the end of the table will make the procedure easier to perform.

2b. The cat's body will need to be hugged close to your body to keep it in the sternal position. *See Figure 4-7.*

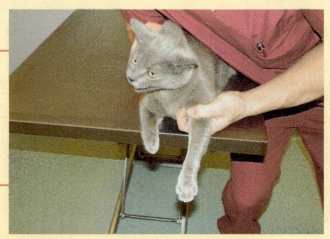

Figure 4-7: Cat with limb extended and vein occluded for cephalic vein venipuncture.

Technical Action

3. The person performing the venipuncture will grasp the left front paw and extend the limb toward them.

Technical Action

4. Occlude the vein by pressing down on the top of the elbow joint with your thumb and then rotate your thumb laterally.

Rationale/Amplification

4a. Your thumb will lay perpendicular to the cat's limb and will be at the bend of the elbow where the radius and humerus meet.

4b. You will be told when to release the vein.

Restraint of the Cat for Jugular Venipuncture

Purpose

- To obtain a large sample of blood quickly

Complications

- Injury to personnel from bites and scratches
- Injury to the cat from excessive force

Equipment

- Examination table

PROCEDURE
Restraint of the Cat for Jugular Venipuncture

Technical Action

1. Place the cat in sternal recumbency with its chest close to the edge of the table.

Rationale/Amplification

1a. This restraint can also be done with the cat sitting on the table if the cat is cooperative this way.

Technical Action

2. Hold the head up, away from the chest. This can be achieved by cupping your hand underneath the jaw and pushing the head up toward the ceiling.

Rationale/Amplification

2a. The person doing the procedure will direct the exact position of the head for maximum visualization of the vein.

Technical Action

3. Your other hand will grasp the front legs and extend them down off the edge of the table.

Rationale/Amplification

3a. The cat's body should be cradled close to your chest. *See Figure 4-8.*

3b. This does not provide the most secure restraint of the body.

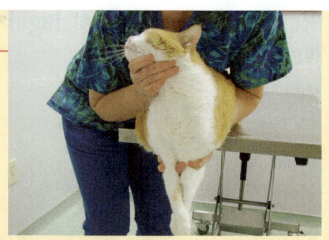

Figure 4-8: The cat is being restrained for jugular vein venipuncture. Note that the forelimbs are extended over the edge of the table. This technique could also be used for a small dog.

Removing a Cat from a Cage

Purpose

To allow the cat to be taken to another area of the hospital for situations such as performance of a physical examination, to undergo treatment, or to be discharged from the hospital.

Complications

- The cat can get loose
- Personnel may be scratched or bitten by the cat as it is being removed from the cage

Equipment

- None required

PROCEDURE
Removing a Cat from a Cage

Technical Action

1. Open the cage door, calling the cat by name.

Rationale/Amplification

1a. Make sure all doors are closed in the room where the cat is housed.

Technical Action

2. Scruff the cat with one hand and lift it up.

Rationale/Amplification

2a. *See Figure 4-9a.*

Figure 4-9a: Cat being scruffed and lifted from a cage.

Technical Action

3. Cradle the cat's abdomen with your other hand and remove the cat from the cage.

Technical Action

4. With the cat still scruffed, place the cat on your hip and close the cage door with your free hand.

Rationale/Amplification

4a. *See Figure 4-9b.*

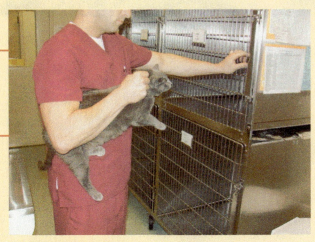

Figure 4-9b: Cat is placed on restrainer's hip while closing the door.

Technical Action

5. Carry the cat close to your body to its destination.

Placing a Cat in a Cage

Purpose

- To house the cat in the hospital or to return a cat to a cage it previously occupied

Complications

- The cat can get loose
- Personnel may be scratched or bitten by the cat as it is being placed in the cage

Equipment

- A cage

PROCEDURE
Placing a Cat in a Cage

Technical Action

1. Carry the cat to the cage by having it scruffed, cradling the body against your hip.

Rationale/Amplification

1a. Make sure all doors are closed in the room where the cat is housed. *See Figure 4-10a.*

Figure 4-10a: Cat scruffed and placed on restrainer's hip.

Technical Action

2. Walk up to the cage and open the door.

Rationale/Amplification

2a. *See Figure 4-10b.*

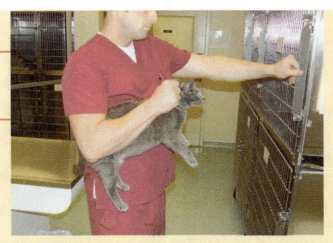

Figure 4-10b: Restrainer opening the cage door.

Technical Action

3. With the cat still scruffed, lift it off your hip and place it as far into the cage as you can with it facing the back wall.

Rationale/Amplification

3a. This gives you some time and distance from the cat when you release it should it decide to turn around in an escape attempt. *See Figure 4-10c.*

Figure 4-10c: Restrainer is placing the cat as far back in the cage as possible with the cat facing the back wall.

Technical Action

4. Quickly release the cat while you hold the cage door with the other hand.

Technical Action

5. Close the cage door.

Rationale/Amplification

5a. Be prepared for the fact that some agitated cats may quickly turn around and attempt to retaliate by swatting or biting. The door needs to be shut quickly.

5b. Make sure the cage is completely latched closed.

Review Questions

1. What type of restraint should always be used with a cat whenever possible?
2. How should a cat be approached prior to restraining it?
3. What warning signs does a cat give to indicate that it is agitated?
4. What is meant by scruffing?
5. Give an example of a restraint technique that could be used so someone could perform a physical examination on a cat.
6. Describe the "cat stretch" restraint technique.
7. How is a cat restrained for a jugular vein venipuncture?
8. Describe how a restrainer occludes the medial saphenous vein.
9. How is a cat removed from a cage?
10. What possible negative reaction might you see when a cat is returned to its cage?

Bibliography

Crow, S., & Walshaw, S. (1997). *Manual of clinical procedures in the dog, cat and rabbit*. Philadelphia: Lippencott-Raven.

McCurnin, D., & Bassert, J. (2006). *Clinical textbook for veterinary technicians* (6th ed.). St. Louis: Elsevier Saunders.

Sheldon, C., Sonsthagen, T., & Topel, J. (2007). *Animal restraint for veterinary professionals*. St. Louis: Mosby Elsevier.

4. To gain control of the cat / restrain the cat.

8. One person restrains while the other uses their thumb to occlude the saphenous veins while also rolling the vein.

Chapter 5

Restraint of Exotic Pets

Each species is a masterpiece, a creation assembled with extreme care and genius.

—Edward O. Wilson

Objectives

- Identify behavioral characteristics unique to each species.
- Describe restraint techniques used on exotic animals.
- Compare and contrast restraint procedures used in each species.
- Identify various facilities, tools, and equipment used in exotic animal restraint.

Key Terms

dyspnea
tachypnea

Restraint of Exotics

A variety of exotic animals are kept as pets. Veterinary technicians and assistants must realize that to the owner, these pets are just as much a companion and family member as the dog or cat. Therefore, understanding the human–animal bond is important with these pets as well. Owners in many cases will be willing to spend money that equals many times the initial cost of the pet to provide care. Owners expect veterinary professionals to be knowledgeable about their particular type of pet. In addition, an owner will expect the clinic personnel to show the same compassion toward their budgie as one would show toward a dog.

A general rule to follow when handling all exotic pets is that you have all the necessary equipment and supplies needed for the procedure ready before the animal is handled as some will not tolerate handling for long.

The following will give information on how to restrain some of the most common exotic pets seen in small animal practice.

Restraint of the Ferret

Ferrets are popular exotic pets so they are typically very accustomed to handling. Their mischievous playful nature, which is maintained throughout their life, makes them endearing pets.

Guidelines for Restraint of the Ferret

The following guidelines will help the technician and assistant in safely handling this species.

- Ferrets are typically very easy to handle and are used to human contact. Their playful, mischievous nature makes them an entertaining pet to own and see in the clinic. Care must be taken when handling them during one of their playful bouts so they do not spring off an examination table or out of a restrainer's hands.
- Some young ferrets just like puppies and kittens may nip and even though it may be during play, it still hurts. One should be careful not to get the ferret's face near their own face. Having a ferret attached to your nose is not pleasant!
- Ferrets can be scruffed just like one would scruff a cat if added control is required.
- When holding a ferret, the hind end of its body should always be supported and not allowed to dangle.

Complications

- Bites to personnel
- Back injury to the ferret

PROCEDURE
Restraint of the Ferret

Technical Action

1. With the ferret on an examination table, floor, or in a cage, grasp it with one hand behind the front limbs.

Rationale/Amplification

1a. If using your right hand, your thumb should be under the ferret's left front leg and your fingers under the right front leg. Alternatively, the ferret can be scruffed just like a cat and lifted. *See Figure 5-1a.*

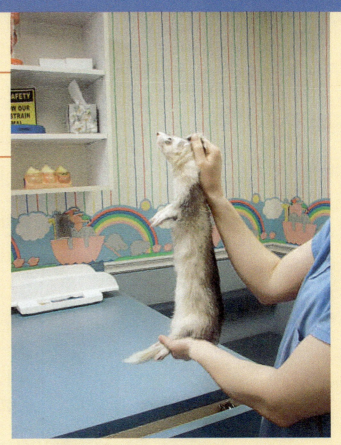

Figure 5-1a: A ferret being restrained by scruffing with support of the hind limbs.

Technical Action

2. Lift up the ferret and use your other hand to support the hind end of the ferret.

Technical Action

3. For someone to perform a physical examination, added restraint may need to be provided, which can be accomplished by scruffing the ferret.

Rationale/Amplification

3a. Most ferrets will immediately yawn right after they are scruffed. The person who is performing the physical examination should take this opportunity to examine the oral cavity. *See Figure 5-1b.*

3b. This restraint can be used to perform such procedures as giving injections, trimming nails, and cleaning ears.

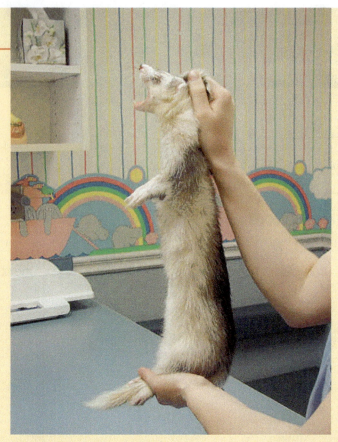

Figure 5-1b: The ferret yawning as it is being scruffed. This is an opportunity to examine the oral cavity.

Restraint of the Rabbit

It is not uncommon to have a veterinary clinic that claims to be strictly a dog–cat practice to be willing to see the occasional rabbit. Rabbits have historically been popular pets due to their beauty and sweet, gentle nature. They come in many different sizes from dwarf to large but all are restrained the same way.

Guidelines for Restraint of the Rabbit

Rabbits are popular pets and are frequently accustomed to handling. They are typically shy and easily startled so one must be calm and quiet when handling this species. The following guidelines should be helpful:

- A rabbit can easily become paralyzed by improper support of the hind limbs when being held. Because of the strength of the hind limbs, they should always be secured so the rabbit cannot kick and injure itself or the handler.

- Covering the rabbit's head so it cannot see is sometimes helpful in reducing the stress that the rabbit is experiencing.

- For a particularly restless rabbit, it can be wrapped in a towel. This "bunny burrito" technique gains control of the limbs so that the rabbit cannot scratch the handler or injure itself from kicks from the hind limbs.

- Rabbits have delicate ears and should never be used as a way to pick one up!
- Rabbits seldom bite but one should never assume that one wouldn't. Occasionally an aggressive rabbit is encountered and they will not shy away from using teeth for defense.

Complications

- Back injury or paralysis to the rabbit due to improper control of the hind limbs during restraint
- Scratches to personnel from the rear claws due to improper control of the hind limbs
- Bites to personnel

PROCEDURE
Restraint of the Rabbit

Technical Action

1. The rabbit should be approached calmly and quietly.

Technical Action

2. The rabbit should be scruffed with one hand and the front end gently lifted.

Rationale/Amplification

2a. If there is not adequate area to scruff and the rabbit is small enough, then the rabbit can be grasped behind its forelimbs.

Technical Action

3. The other hand should immediately reach under the hind limbs and hold them.

Rationale/Amplification

3a. This prevents the rabbit from kicking. *See Figure 5-2a.*

3b. The hind limbs should never be allowed to dangle.

Figure 5-2a: A rabbit being held so that its hind limbs are secured.

Technical Action

4. If the rabbit needs to be carried to another location, the hand that is scruffing the rabbit can be tucked under the opposite arm at the elbow. The rabbit's body is rested on the arm, with the other hand under it holding the rear limbs.

Rationale/Amplification

4a. This covers the rabbit's head, thus blocking its view. *See Figure 5-2a.*

Technical Action

5. A rabbit can be wrapped in a towel to create a "bunny burrito." This is particularly helpful for very frightened, uncooperative rabbits.

Rationale/Amplification

5a. This technique gains control of the limbs. *See Figure 5-2b.*

5b. If returning a rabbit to a cage, the rear end should always be placed in the cage first before the front end is released.

Figure 5-2b: A rabbit wrapped in a towel, creating a "bunny burrito."

Restraint of the Pet Bird

Many different types of birds are kept as pets. While small birds such as finches and canaries are kept as pets, these are not as commonly seen in practice as the different types of parrots. They range in size from the budgie to the macaw. Many represent to owners not only an emotional investment but a monetary one as well. Because some parrots such as the macaws are quite large, they can inflict serious bites. It may take some practice to obtain confidence when handling them. It is best to practice handling small birds such as the budgie or cockatiel before attempting to handle the larger types. Regardless of the type of bird, the restraint technique is the same.

Guidelines for Restraint of the Pet Bird

- Although parrots can be very tame, they can inflict injury with their beaks and claws when restrained for procedures. Care must be taken that the head is controlled to avoid being bitten. Even a small bird like the budgie can inflict painful bites. The extent of injury possible with a large parrot goes without saying.

- Birds have a tremendous amount of strength in their claws and care needs to be taken as to not have hands in the way of being grasped.
- Always ask the owner if the bird's wings are trimmed. Realize that trimmed wings do not necessarily prevent flight. In many cases, it reduces flying ability rather than completely eliminating it.
- If a bird is in a cage when restraint is attempted, remove all toys and any other objects that might get in the way of trying to grasp the bird.
- If the bird is on the owner's shoulder, hand, or arm, always have them put the bird on the examination table. Never take the bird from the owner as injury to the owner or the restrainer may result, as birds can be very territorial about their owners.
- The door(s) to the room in which the restraint is taking place must be closed. Ideally, restraint of birds should be performed behind two closed doors to provide added security against escape.
- The bird needs to be held in such a way that the head and specifically the mandible is secured. This can be done with or without a towel depending on the handler's level of experience. Leather gloves should never be used as it terrifies the bird and sensitivity to the degree of pressure applied to the body is lost.
- If a towel is used to catch and restrain a bird, a paper towel is adequate for a budgie or a finch, whereas a bath towel may be needed for a macaw. The towel serves more as a distracter for the bird and a confidence builder for the restrainer rather than protection for the hands!
- The bird lacks a diaphragm and utilizes the free movement of its sternum to breathe. During restraint, the bird's chest should not be compressed, as this will cause **dyspnea.**
- If at any point the bird becomes stressed to the point of **tachypnea or dyspnea**, the bird should be released immediately. Many birds have died just from the sheer act of restraint. This is always a possibility if the bird is ill upon presentation at the hospital. Some birds may be too ill to handle at all.

Complications

- Bites and scratches to personnel
- Respiratory distress due to stress from restraint or improper restraint
- Death to the bird

PROCEDURE
Restraint of the Pet Bird

Technical Action

1. The door to the room should be closed prior to restraint being attempted. If the restraint is occurring in a treatment room, clinic employees need to be made aware of what is going on so no one inadvertently opens the door. Ideally the handling of a bird should happen behind two closed doors to eliminate the possibility of escape.

Technical Action

2. If the bird is in its cage, remove all toys and other items that can get in the way of catching the bird. If the bird is on the owner's shoulder or other body parts, ask the owner to place the bird on the table.

Rationale/Amplification

2a. In some situations, the cage will have barely enough room for your hand, let alone trying to dodge objects.

2b. When trying to catch a bird in a cage, always try to grasp it when its beak is attached to a cage bar. Some birds, in an attempt to prevent being caught, will lay on their backs in a defensive posture and use their claws to protect themselves. Wait until the parrot gets up and try again.

2c. The bird may fly away when placed on the examination table. Ultimately it may need to be caught on the floor.

Technical Action

3. No matter where the bird is when you are attempting to catch it, you need to come behind it and grasp it right behind and below the head and immediately grasp the body securing the wings to the torso with the other hand. This technique can be performed with or without a towel.

Rationale/Amplification

3a. The goal is to grasp the bird either at the bottom of the mandible or just below it so the head is secure. The wings are secured so they do not get injured from flapping. The bird should never be held by the head alone. *See Figure 5-3a.*

Figure 5-3a: Proper restraint of a parrot. Note that the head is secured, the body is upright, and there is no pressure on the chest.

Technical Action

4. The bird should be held upright without putting pressure on the chest.

Rationale/Amplification

4a. Birds require the movement of their sternum to breathe. *See Figure 5-3b.*

Figure 5-3b: Restraint of the parrot using a towel.

Restraint of the Guinea Pig

Guinea pigs are very popular due to their small size and gentle personalities. They rarely bite and many love to be cuddled.

Guidelines for Restraint of the Guinea Pig

- Although typically friendly, guinea pigs have a tendency to startle easily. They should be approached calmly and quietly but assertively.
- They can be a bit difficult to catch in a cage as they may scurry to try to get away.
- Guinea pigs are very vocal and may squeal or whistle when they are being caught or restrained.
- They should be held with two hands, one around the chest and the other supporting the hind limbs.
- If a guinea pig is being examined on an examination table, it helps to have a towel on the table for traction. Always have one hand on the pet at all times to prevent it from jumping off the table.
- Guinea pigs seldom bite but one can get scratched by their rear claws.

Complications

- Scratches to personnel
- Bites to personnel

PROCEDURE
Restraint of the Guinea Pig

Technical Action

1. To restrain a guinea pig, it first must be caught. Whether from a cage or an examination table, one hand should grasp the guinea pig at the shoulders. While gently lifting the hind limbs can be cupped with the other hand.

Figure 5-4a: Restraint of the guinea pig.

Rationale/Amplification

1a. Make sure the guinea pig is grasped gently but securely so it does not escape and fall to the floor. See Figure 5-4a.

Technical Action

2. Once held, the hand holding the shoulders can be shifted so that the ventral chest is cupped. The hind limbs are held with the other hand.

Rationale/Amplification

2a. *See Figure 5-4b.*

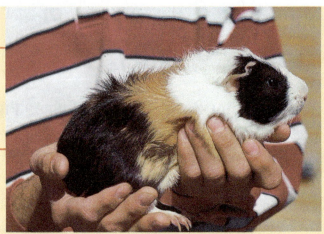

Figure 5-4b: This guinea pig is being restrained with the forelimbs held in the left hand and its hind limbs secured with the right. Alternatively, to provide more control for a more restless guinea pig, the left hand can be used to restrain the guinea pig by placing the hand over the shoulders. In this case, your hand will be over the top of the guinea pig with the index finger and thumb directly behind the forelimbs. The right hand will support the hind limbs as shown.

Review Questions

1. In what way can a restrainer get injured by a ferret?
2. What is the best way to examine the oral cavity of a ferret?
3. When handling rabbits, what is the number one concern with regard to restraint?
4. How is a towel used in restraining a rabbit?
5. How can a parrot inflict injury on a restrainer?
6. What potential danger does restraint pose for a bird?
7. Why is compression of the chest in a bird dangerous during restraint?
8. Why are leather gloves inappropriate for restraint of a bird?
9. What characteristics make guinea pigs desirable pets?
10. In what way might a restrainer get injured by a guinea pig?

Bibliography

Crow, S., & Walshaw, S. (1997). *Manual of clinical procedures in the dog, cat and rabbit*. Philadelphia: Lippencott-Raven.

McCurnin, D., & Bassert, J. (2006). *Clinical textbook for veterinary technicians* (6th ed.). St. Louis: Elsevier Saunders.

Quesenberry, K. D. V. M., & Carpenter, J. D. V. M. (2008). *Ferrets, rabbits and rodents* (3rd ed.). St. Louis: Saunders.

Sheldon, C., Sonsthagen, T., & Topel, J. (2007). *Animal restraint for veterinary professionals*. St. Louis: Mosby Elsevier.

Unit 2

Large Animal Restraint

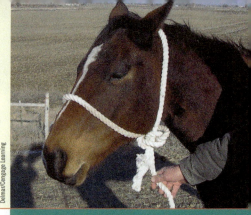

Chapter 6

Ropes and Knots

Objectives

- Identify the types of ropes available and explain how to maintain them.
- List and describe the types of ropes used in large animal medicine.
- Describe how to tie basic knots used in large animal medicine.
- Describe how to build a rope halter for cattle and sheep.

Key Terms

bight
hondo
lariat
loop
sisal
tensile strength

The greatness of a nation and its moral progress can be judged by the way its animals are treated.

—Mahatma Gandhi

Ropes

For years, ropes were made out of hemp, and although very durable and weather resistant, their roughness caused rope burns and irritation. Ropes these days are generally made of synthetic materials, cotton, or **sisal**. Synthetic ropes are also weather resistant and difficult to break, but just like hemp, they tend to burn either the patient or the handler if pulled rapidly across the skin. An added benefit is that these synthetic ropes tend to be of lighter weight than cotton or sisal. Cotton ropes most often are used for positioning large animals, because they are softer and less apt to burn. They should not be left out in the sun or inclement weather, however, because they will break down quickly and lose **tensile strength**. Specialized ropes, such as **lariats** (commonly used to rope a calf), may be coated with wax to help them slide through the **hondo** (or **loop**) more effectively. The diameter and coating of lariats tend to cause rope burns when used for restraint purposes. All ropes can be purchased in almost any length and thickness at the local hardware or ranch supply store.

SAFETY ALERT

- All ropes, halters, and lead ropes used for restraint should be checked often for signs of fraying or weakness.
- Nylon halters and ropes should be washed regularly to lessen the spread of disease.
- Washing cotton ropes speeds their degradation.
- Lariats are not meant to be washed and will lose their special handling characteristics if that is done.

Finishing the End of a Rope

Whenever a rope is cut, it is important to do something with the raw end to prevent it from unraveling. The simplest method is to tie a knot in the end, but that makes the end bulky and difficult to manipulate. Whatever method you choose, make sure it is appropriate for the intended use of the rope.

PROCEDURE
Finishing or Securing the End of a Rope

Technical Action

1. Tape the end.

Rationale/Amplification

1a. Using black electrician's tape, start wrapping the rope about 1 cm (½ inch) from the end with the tape.

1b. Pull very hard, hard enough to compress the strands of the rope.

1c. Continue up the rope for approximately 5 cm (2 inches).

Technical Action

2. Whip the end.

Rationale/Amplification

2a. Select a thin piece of nylon string or cord about 30 cm (12 inches) long.

Rationale/Amplification

2b. Lay the cord on the rope in a U-shape, with the bottom of the U about 5 cm (2 inches) from the end of the rope and the arms of the U heading off the end of the rope (*see Figure 6-1a*).

2c. Hold the loop of cord against the rope.

2d. Using the long arm of the cord, and starting about 1 cm from the end of the rope, tightly wrap the rope with the cord (*see Figure 6-1b*).

2e. When at least 2.5 cm (1 inch) of the rope is covered, pass the long end of the cord through the loop (*see Figure 6-1c*).

2f. Pull the short end of the cord until the loop disappears and the long end of the cord is sucked up tight to the wraps (*see Figure 6-1d*).

2g. Cut both ends of the cord close to the wrap.

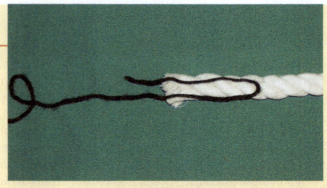

Figure 6-1a: Step one to whipping the end of a rope.

Figure 6-1b: Tightly wrapping the end of the rope with the cord.

Figure 6-1c: Passing the end of the whipping cord through the loop.

Figure 6-1d: Pull the end of the cord in the direction shown.

Technical Action

3. Burn the end.

Rationale/Amplification

3a. Done with a nylon rope.

3b. Hold a lighter flame to the raw end of the rope until the strands melt and curl back.

3c. Rotate the rope to get even coverage.

3d. Takes roughly 30 seconds to get complete and even melting up to 1 cm up the rope.

3e. Tape may then be applied as above for a finished appearance.

Quick-Release Knot

This simple, easy-to-tie knot is used most frequently by horse people. The main drawback to this knot is that it continues to tighten when the long end of the rope is pulled on, which can sometimes make it difficult to release.

Purpose

- To tie a horse or llama to a hitching post, rail, or hook
- To allow the lead rope to be untied quickly and easily by simply pulling on the free end
- To secure any restraint rope that may need to be released quickly

PROCEDURE
Tying a Quick-release Knot

Technical Action

1. Pass the rope over the hook or rail from left to right.

Rationale/Amplification

1a. Make the short end at least 45–60 cm (18–24 inches) long.

Technical Action

2. Make a loop in the short end and pass it over the long end (*see Figure 6-2a*).

Rationale/Amplification

2a. The loop should be about 10–15 cm (4–6 inches) in diameter.

Figure 6-2a: Loop in the short end held over the long end or the rope attached to the animal's halter.

Technical Action

3. Pass the short end behind the long end and the loop and push a **bight** (a fold of rope) into the loop (*see Figure 6-2b*).

Rationale/Amplification

3a. Make the bight at least 25 cm (10 inches) long.

Figure 6-2b: Placing the bight within the loop.

4. Pull the bight to tighten the knot.

4a. Animals learn quickly to untie this knot by pulling the short end, so you can pass the short end through the bight.

Bowline

The bowline is probably the first knot learned by those who work around boats and water, because no matter how much force is pulled against it, it can still be untied. It's named such because this knot has been used for centuries to secure the tug line to the bow of a boat.

Purpose

- Used to make a fixed-diameter loop
- Easily released even when pulled extremely tight, so used for securing an animal to a post or for dragging a dead animal

PROCEDURE
Tying a Bowline

Make a loop in the long end of the rope such that the short end of the rope overlaps the long end (*see Figure 6-3a*).

Figure 6-3a: Note carefully how this loop was formed. It is important that the short end cross over the long end.

Technical Action

2. Pass the short end of the rope up through the loop.

Rationale/Amplification

2a. This is the rabbit coming out of the hole.

Technical Action

3. Reach under the long end of the rope and grasp the short end such that it wraps around the long end (*see Figure 6-3b*).

Rationale/Amplification

3a. This is the rabbit running around the tree.

Figure 6-3b: Wrapping the short end of rope around the long end.

Technical Action

4. Pass the short end of the rope back through the loop in the opposite direction of the first pass (*see Figure 6-3c*).

Rationale/Amplification

4a. This is the rabbit running back down into the hole.

Figure 6-3c: Pass the short end through the loop, and pull in the direction indicated by the arrow.

Technical Action

5. Tighten the knot by pulling on both the long and short ends.

Tomfool Knot

Named for the way the knot disappears if the free ends are pulled, the tomfool knot is handy to know when you need to secure or hobble the feet of an animal. Tied correctly, the knot will hold fast without cutting off the circulation, even if the animal struggles against it.

Purpose

• Used to tie two legs together

PROCEDURE
Tying a Tomfool Knot

Technical Action

1. Grasp the center of the rope in both hands.

Rationale/Amplification

1a. The right hand should be positioned thumb up and the left hand thumb down.

Technical Action

2. Rotate both hands counterclockwise to form two loops.

Rationale/Amplification

2a. Now both thumbs are in the middle facing each other (*see Figure 6-4a*).

Figure 6-4a: Note how hands are positioned in relation to each other and to the rope.

Technical Action

3. Bring the loops together so that the right loop overlaps the left loop.

Rationale/Amplification

3a. Overlap by half so that one side of the right loop is in the middle of the left loop.

Technical Action

4. Pull the side of the right loop through the left loop and the near side of the left loop through and over the right loop.

Rationale/Amplification

4a. *See Figure 6-4b.*

Figure 6-4b: Passing the two loops through each other.

Technical Action

5. Pull both hands apart to create two loops knotted together in the middle (*see Figure 6-4c*).

Rationale/Amplification

5a. If you pull on the ends of the rope when the loops are not over something, the whole thing will come undone.

Figure 6-4c: Pull in direction indicated by arrows.

Technical Action

6. Place each loop over a limb, and pull each end very tight to snug the loops around the legs.

Rationale/Amplification

6a. Finish off with a simple overhand or slip knot to keep the animal from wiggling loose.

Double Half Hitch Knot

Also known as a clove hitch. People who work regularly around livestock use this hitch all the time; often tying it with little or no thought.

Purpose

- To secure a rope to a post, rail, or hook when it does not need to be released quickly
- May be used to tie a horse, cow, or llama to an object, although this is not recommended for safety reasons

PROCEDURE
Tying a Double Half Hitch

Technical Action

1. Pass the rope around the post.

Rationale/Amplification

1a. May be a rail, hook, or whatever you wish to tie to.

Technical Action

2. Pass the short end under the long end and then back over the top.

Rationale/Amplification

2a. In effect, you are creating a closed loop around the post.

Technical Action

3. Continue down between the post and the loop you just formed.

Rationale/Amplification

3a. *See Figure 6-5a.*

Figure 6-5a: Pull in direction shown by arrow.

Technical Action

4. Pull it tight.

Technical Action

5. Pass the short end over and under the long end, forming a loop.

Technical Action

6. Pass the short end up through the loop and pull it tight.

Rationale/Amplification

6a. *See Figure 6-5b.*

Figure 6-5b: Second half of the double half hitch.

Tail Tie

Ropes often are tied to a horse's tail to help us maneuver the animal more effectively. Horses have very strong tails that can support their body weight. Cattle, on the other hand, have very weak tails that may be broken or even pulled off if tied by the tail.

Purpose

- Used to lift or move the back end of a recumbent or ataxic horse
- Used in horses to link one horse to another, head to tail (as in a pack string)
- Used in horses to tie the tail out of the way
- May be used in cattle *only* to hold the tail out of the way, *never* to lift or move the animal

PROCEDURE
Placing a Tail Tie

Technical Action

1. Lay a rope over the tail at the tip of the tail bone.

Rationale/Amplification

1a. Make the short end about 18 inches long.

Technical Action

2. Fold all the tail hairs up over the rope.

Rationale/Amplification

2a. This can be difficult in a horse with a very short, thin tail.

Technical Action

3. Pass the short end of the rope behind the tail, and make a fold or bight in it.

Technical Action

4. Pass the fold or bight over the folded tail and under the rope, which is looped around the tail.

Rationale/Amplification

4a. *See Figure 6-6a.*

Figure 6-6a: Passing the bight through the tail loop.

Technical Action

5. Pull tight.

Rationale/Amplification

5a. *See Figure 6-6b.*

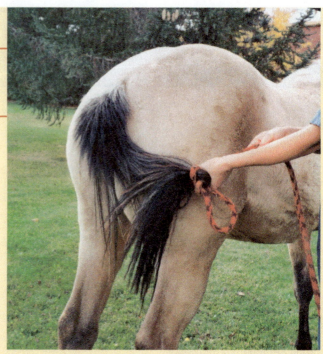

Figure 6-6b: Finished tail tie.

Braiding an Eye Splice

Placing an eye splice at the end of a rope makes it useful for many situations requiring a slip loop. Once having accomplished the braiding technique, you will be able to splice ropes together or finish off the end of the rope by braiding back.

Purpose

- To create a permanent loop in the end of a rope that can withstand a great deal of force
- Basis for creating a hondo in a lariat

PROCEDURE
Braiding an Eye Splice

Technical Action

1. Unravel 8–10 inches of the rope.

Technical Action

2. Make a bight or loop such that the unraveled ends are at right angles to the axis of the still-braided strands of the rope.

Rationale/Amplification

2a. Make the bight (eye) whatever size suits your needs (*see Figure 6-7a*).

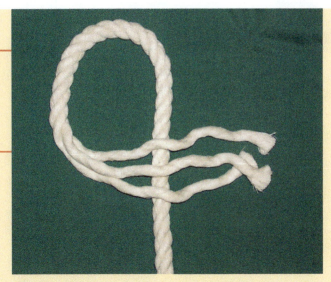

Figure 6-7a: First step in creating an eye splice.

Technical Action

3. Lift one strand of the intact rope, and pass the center strand of the loose ends beneath it. This will be strand 1.

Rationale/Amplification

3a. *See Figure 6-7b.*

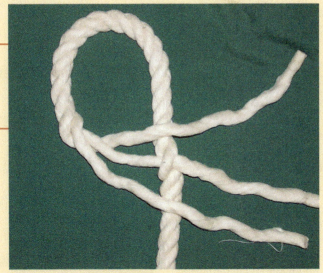

Figure 6-7b: Passing the first strand of rope through the rope.

Technical Action

4. Raise the strand in the side of the rope next to where strand 1 exits.

Technical Action

5. Pass loose strand 2 beneath this strand, so that strand 2 enters the rope where strand 1 exits.

Rationale/Amplification

5a. *See Figure 6-7c.*

Figure 6-7c: Passing the second strand of rope through the rope.

Technical Action

6. Turn the ropes over and pass loose strand 3 under the strand where strand 2 exits, and it exits the rope where strand 1 entered.

Technical Action

7. Continue passing the loose strands over each other and through the rope until their ends are reached or cut.

Rationale/Amplification

7a. *See Figure 6-7d.*

Figure 6-7d: Finished product. Ends may be trimmed to smooth the appearance of the rope.

Rope Halters

In large animal medicine, it is common to make your own halters, especially for cattle. Building your own halter not only saves you money but makes it possible to have on hand halters that are a variety of weights and sizes. Horse halters are more complicated to build and, in this author's experience, not suitable for most of the restraint procedures. Knowing how to build a temporary rope halter, however, is extremely useful when you need to restrain a horse that has been injured and has no halter immediately available.

Purpose

- Restraint or control of the heads of sheep, goats, or cattle
- Leading cattle, sheep, goats, and horses
- Tying cattle, sheep, or goats to a fixed object

Equipment

- 12–14 feet of three-strand cotton or nylon rope ½-inch thick for adult cattle, ⅜-inch thick for sheep, goats, and small calves
- 10–14 feet of ½-inch cotton or nylon rope for a horse
- Electrical tape, hog rings, or cord to secure the end of the rope

PROCEDURE
Building a Rope Halter for Cattle or Sheep

Technical Action

1. Finish one end of a 12-foot to 14-foot three-strand cotton or nylon rope.

Rationale/Amplification

1a. Wrap end tightly with electrician's tape or whip the end (refer to "Procedure for Finishing or Securing the End of a Rope").

Technical Action

2. Make the nosepiece for a cow.

Rationale/Amplification

2a. Measure 18 inches from the finished end of the rope for a large breed of cow (12–14 inches for a smaller breed or a calf).

2b. At this 18-inch point, separate the strands enough to insert the long end (*see Figure 6-8a*).

2c. Have two strands on top and one underneath.

2d. Insert the long (unfinished) end so that the short (finished) end points down and the long end points up.

Figure 6-8a: First step in creating the loop for the nosepiece.

Technical Action

3. To make the nosepiece for a sheep: Complete the directions above for this step.

Rationale/Amplification

3a. Measure 6–8 inches using ⅜-inch rope.

Technical Action

4. About 4 inches from the loop, separate the strands of the long end. Place the short, finished end through the long end.

Rationale/Amplification

4a. Two strands on top and one underneath (*see Figure 6-8b*).

Figure 6-8b: Passing the short end of rope through the strands separated on the long end.

Technical Action

5. Pull the short end all the way through, so that a single loop is formed with four strands on top and two strands underneath where the ends intertwine.

Rationale/Amplification

5a. *See Figure 6-8c.*

Figure 6-8c: Pull the short end tightly to secure the loop.

Technical Action

6. To form the crown, follow steps 7 through 10.

Rationale/Amplification

6a. The piece that goes behind the ears of the cow.

Technical Action

7. Take the short end of the rope in your left hand. Place your right hand about 3 inches from your left hand.

Technical Action

8. Turn your left hand toward you and your right hand away from you so that the rope twists open.

Technical Action

9. Continue to twist the rope until three small loops are formed by the strands.

Rationale/Amplification

9a. *See Figure 6-8d.*

Figure 6-8d: These loops were created simply by twisting the rope in a direction opposite to its existing twist.

Technical Action

10. Place the long end of the rope through these three loops.

Rationale/Amplification

10a. Pull all but about 2 feet of the long end of the rope through the loops.

10b. For sheep, the crown needs to be only about 12 inches long.

Technical Action

11. Now place the long end through the loop completed in steps 1 through 4.

Rationale/Amplification

11a. This will go under the jaw of the sheep or cow and tightens when you pull on the lead.

Technical Action

12. Adjust the halter so it looks like the one in *Figure 6-8e* and finish the lead-rope end as described in Procedure for Finishing or Securing the End of a Rope.

Figure 6-8e: Finished product.
The crown length is adjustable to fit a variety of head sizes.

PROCEDURE
Building a Temporary Rope Halter for a Horse

Technical Action

1. Loop the rope around the horse's neck.

Rationale/Amplification

1a. Rope should be loose enough to fit your hand perpendicularly with ease between the rope and the horse's throat latch.

Technical Action

2. Tie a bowline to secure the loop.

Rationale/Amplification

2a. Make sure it is not a slip knot (*see Figure 6-9a*).

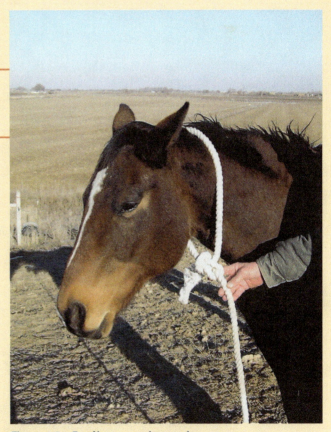

Figure 6-9a: Bowline secured around horse's neck.

Technical Action

3. Fold the long end of the rope up through the neck loop.

Rationale/Amplification

3a. *See Figure 6-9b.*

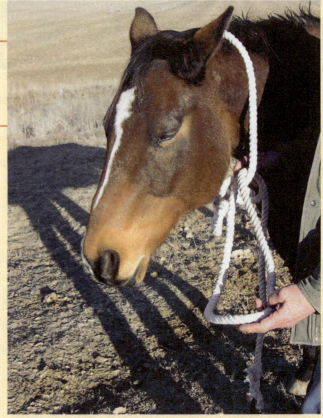

Figure 6-9b: Passing a bight or fold of rope through the neck loop.

Technical Action

4. Pass the bight or loop over the bridge of the horse's nose.

Rationale/Amplification

4a. *See Figure 6-9c.*

Figure 6-9c: Finished product.

Technical Action

5. Secure the second loop by tying a second knot at the throat latch.

Rationale/Amplification

5a. For a temporary halter, just tie a simple overhand knot.

Review Questions

1. What are the advantages and disadvantages of synthetic rope?
2. How does a lariat differ from other restraint ropes?
3. Why is it important to check ropes frequently?
4. What is a quick-release knot most often used for?
5. How can a quick-release knot be modified so that the animal doesn't release itself?
6. What are the advantages of using a bowline?
7. Why does a tail tie have limited use in cattle?
8. What is a tomfool knot and what is it used for?
9. Name two ways to finish off the end of a rope.
10. What else can the braiding back technique be used for?

Bibliography

Heersche, G. (1982). *How to make a rope halter*. Lexington, KY University of Kentucky College of Agriculture Cooperative Extension Service, 4 AA-0-200.

Leahy, J., & Barrow, P. (1953). *Restraint of animals* (2nd ed.). Ithaca, NY: Cornell Campus Store.

Chapter 7

Restraint of the Horse

Objectives

- Identify behavioral characteristics unique to the horse.
- Describe restraint techniques used on the horse.
- Identify various facilities, tools, and equipment used in restraint of the horse.

Key Terms

halter
stock
twitch

How do you catch a loose horse? Make a noise like a carrot.

—British Cavalry Joke

Restraint of the Horse

Horses are the large animal most accustomed to handling. Whether it involves haltering, grooming, or transportation, frequent handling is a component of all equine husbandry protocols.

Guidelines for Restraint of the Horse

The ability to work cooperatively with horses takes many years to develop. The following guidelines, however, will help the inexperienced handler avoid many problems:

- Horses are large herd animals. Their general approach is to run first and ask questions later. Thus, most accidents involving equines are the result of nervous, not aggressive, behavior on the horse's part. Having said that, horses that choose to do so can inflict extensive damage via biting, kicking (horses most often kick caudally), and striking (using the front limbs).
- Always let horses know that you are in the area. Horses do not appreciate surprises. Speak calmly; avoid loud noises and sudden movement.
- When standing near a horse, place your hand gently on the animal. This will alert you to any pending movement on the horse's part.
- Never go under a horse.
- Never stand directly behind in the blind spot of a horse.
- If you must walk behind the horse, walk very close to the horse with your body touching the animal. Alternatively, you can walk behind at a distance of 15 feet from the hind end.
- Do not permit yourself to be pinned or squeezed between the horse and a solid object. For example, do not stand between the horse and the wall. Change positions such that the horse is between you and the wall.
- Horses occasionally will cow kick (kick cranially with a hind limb) or strike using a forelimb.
- Most handler accidents occur when the horse is placed in a confined area (**stock**, trailer) or is within physical contact distance of another horse. Pay close attention in these situations.

Horse Behavior

Unmanageable situations can often be avoided by correctly reading the emotions of a horse. Ears pinned backward often indicate anger or a warning. These signs warn handlers that they may be bitten or kicked. Ears forward show interest or suspicion. Eyes and nostrils also show emotion and reflect temperament. Dilated nostrils reflect interest, curiosity, or apprehension. When the eyes flash, nostrils dilate, and muscles tense, the horse is likely about to react. *Figure 7-1* outlines some of the common cues to look for in a horse's behavior.

Rules of Tying

To facilitate safety of both horses and handlers, the following guidelines should be adhered to when tying horses:

- Always use a quick-release knot.
- Secure to solid post or hitching rack. Never tie to gates, fence rails, or any object that a frightened animal could pull apart.

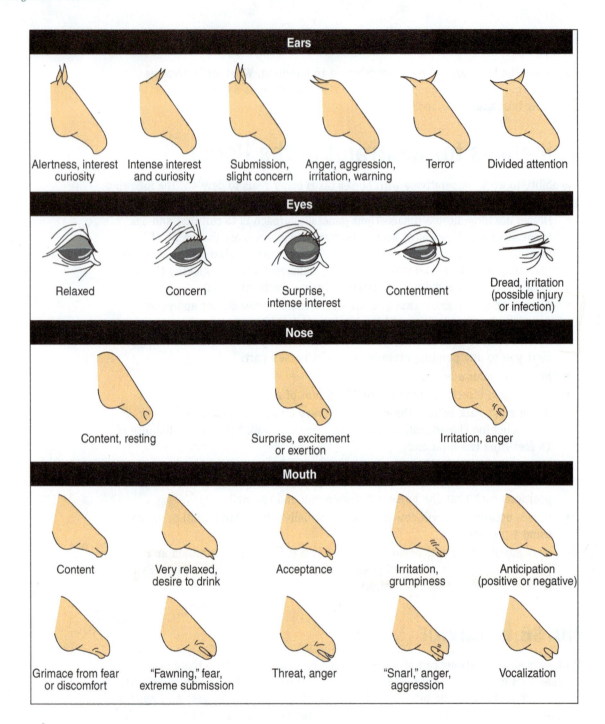

Figure 7-1: Common cues to horse behavior.

- Lead ropes should be tied short and high enough to prevent animal from lifting a leg over the rope. This is usually at the level of the animal's head.
- Ensure a minimum of 12 feet between animals.

Stock

Purpose

- Inhibit horse's movement through bodily confinement

Complications

- Injury to horse
- Bodily injury to personnel

Equipment

- Stock
- **Halter** and lead rope

PROCEDURE
Using Stock

Technical Action

1. Place halter and lead rope on horse.

Technical Action

2. Open front and back gates of stock.

Rationale/Amplification

2a. *See Figure 7-2.*

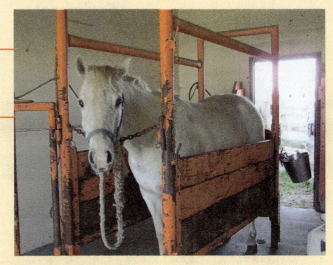

Figure 7-2: Equine stock.

Technical Action

3. Walk horse into stock while you remain outside just to the left of the stock.

Rationale/Amplification

3a. If the horse refuses to enter, the handler can precede the horse into the stock. Be certain that the front of the stock is open before attempting this.

3b. Do not look at the horse as you approach the stock. Keep looking forward.

3c. Many horses will halt 3 feet from the stock. Allow them to quietly assess the situation for 2–3 minutes, then cluck to encourage forward movement.

Technical Action

4. Close front gate, close hind gate. Tie the horse at this time if warranted.

Rationale/Amplification

4a. Horses should never be placed in cattle chutes.

4b. Never leave a horse unattended in a stock.

Haltering and Leading

Purpose

- To provide fundamental restraint for horse

Complications

- Injury to handler if stepped on by the horse

Equipment

- Halter
- Lead rope

PROCEDURE
Haltering the Horse

Technical Action

1. Approach horse slowly from the left side in the area of the shoulder.

Rationale/Amplification

1a. The shoulder is the safest area of the horse.

1b. If the horse is not aware of your presence, make a soft sound that will alert him.

1c. Never approach a horse directly from behind.

1d. Most horses are accustomed to being handled from the left side.

Technical Action

2. Place lead rope around neck.

Rationale/Amplification

2a. The lead rope should be as close to the head as possible, not low on the neck.

2b. The lead rope should be attached to the **ventral** D-ring of the halter.

Technical Action

3. Place nose band over horse's nose and buckle strap behind ears.

Rationale/Amplification

3a. Some halters have a ventral clasp that is buckled instead of the crown piece, which goes behind the ears.

3b. Try to minimize contact with the ears, because many horses are sensitive about their ears being touched.

3c. See *Figure 7-3*.

Figure 7-3: Haltering the horse.

PROCEDURE
Leading the Horse

Technical Action

1. The horse should be haltered with the lead rope attached.

Rationale/Amplification

1a. See *Figure 7-4*.

Figure 7-4: Leading the horse.

Technical Action

2. Stand on the left side of the horse.

Rationale/Amplification

2a. The left side is also referred to as the near side; the right side is called the off side.

Technical Action

3. Hold lead rope in right hand 12 inches from the halter. Coil the excess rope and place in left hand.

Rationale/Amplification

3a. Never wrap any portion of a lead rope around yourself.

3b. If needed, the handler can always grasp the cheek piece or chin strap of the halter.

Technical Action

4. Walk forward at a brisk pace.

Rationale/Amplification

4a. Do not look at the horse. He knows that he should be following you.

Technical Action

5. Should the horse become unruly, circle to the left. This will allow you to act as a pivot while the horse circles around you.

Applying Chains

 ### Purpose
- Increase the amount of restraint through increased pressure

Complications
- Head tossing
- Tissue trauma (buccal or oral)

Equipment

- Halter
- Lead rope with shank chain

PROCEDURE
Applying Chains

Technical Action

1. Halter the horse.

Rationale/Amplification

1a. Chains can be used over the nose, under the chin, in the mouth, or under the lip, if necessary. All of the methods increase the amount of restraint, with varying degrees of discomfort.

1b. Horses should never be tied using a chain lead rope, because it could result in severe trauma if the horse were to pull back.

Technical Action

2. Chain over nose:
Disconnect chain from ventral halter D-ring, and pass chain through left side ring, over nose, and attach to right side ring.

Rationale/Amplification

2a. If the chain is long enough, it can be passed through the right-side D-ring and clipped to the ventral ring, or it can be passed through the right-side D-ring and clasped to the right-cheek D-ring.

2b. Looping the chain one time around the halter's nose band will help prevent the chain from slipping down the nose. *See Figure 7-5.*

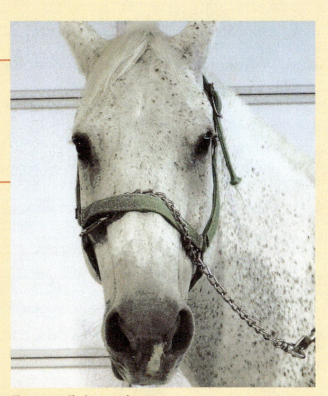

Figure 7-5: Chain over the nose.

Technical Action

3. Chain under chin:

Pass chain through left side D-ring under chin and attach to right side D-ring.

Rationale/Amplification

3a. *See Figure 7-6.*

Figure 7-6: Chain under the chin.

Technical Action

4. Chain in mouth:

a. Pass chain through left side D-ring under chin and attach to right side D-ring.
Loosen chain such that there is enough slack to reach horse's lips.

b. Place right arm ventral to head and grasp head in the area of nose band with right hand.

c. Insert left thumb in commissure of the lips. As horse opens mouth, slide chain in (as if placing a bit in mouth).

d. Apply slight pressure to chain to remove slack.

Rationale/Amplification

4a. The chain should rest comfortably, just touching the commissure on the lips. If the chain is too tight or too low in the mouth, it will cause unnecessary discomfort.

4b. Placing a chain in the mouth is rarely done in veterinary practice. Care should be taken to prevent harm to the delicate tissues of the lips.

4c. *See Figure 7-7.*

Figure 7-7: Chain in the mouth.

Technical Action

5. Chain under lip:

a. See steps 4a and 4b above.

b. Lift upper lip and slide chain over gum under the lip.

c. Apply slight pressure to remove slack.

Rationale/Amplification

5a. This can be extremely painful when pressure is applied. Do not maintain pressure when the horse is cooperative. The chain should rest snugly without causing any discomfort.

Twitches

Purpose

- Application of pressure with the intent to distract attention or induce endorphin release

Complications

- Personnel injury
- Trauma to lip

Equipment

- Halter
- Lead rope
- **Twitch**

PROCEDURE
Hand Twitching

Technical Action

1. Apply halter and lead rope.

Rationale/Amplification

1a. Never attempt to apply a twitch without haltering the horse.

1b. Twitches provide minor pain for the horse, thereby creating a distraction during a clinical procedure.

1c. If applied to the nose, twitches are thought to cause endorphin release, thereby suppressing pain in the horse.

Technical Action

2. Turn head toward you and grasp the loose skin in the neck area just **cranial** to the shoulder. Grasp a large amount of skin, twist slightly, and hold firmly.

Rationale/Amplification

2a. Hand twitching is performed most commonly in the neck area. A twitch should never be applied to a horse's ear. Holding onto the nose manually is difficult and better accomplished using a mechanical twitch.

2b. Turning the head toward you loosens the skin, making it easier to grasp.

2c. *See Figure 7-8.*

Figure 7-8: Applying a hand twitch to the neck.

PROCEDURE
Applying Mechanical Twitches

Technical Action

1. Apply halter and lead rope.

Rationale/Amplification

1a. Never attempt to apply a twitch without haltering the horse.

1b. Twitches provide minor pain for the horse, thereby creating a distraction during a clinical procedure.

1c. If applied to the nose, twitches are thought to cause endorphin release, thereby suppressing pain in the horse.

Technical Action

2. Select twitch type.

Rationale/Amplification

2a. Mechanical twitches include the Kendal humane twitch, chain twitch, or rope twitch.

2b. The Kendal humane twitch is hinged, which prevents over tightening. It can also be made self-retaining.

Technical Action

3. Place hand through twitch and firmly grasp upper lip.

Rationale/Amplification

3a. The horse will most likely resist by raising or flipping its head up and down.

3b. Grasp as much of the upper lip as possible.

3c. *See Figure 7-9.*

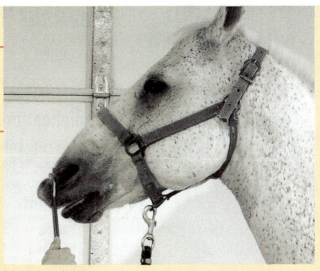

Figure 7-9: Applying a mechanical twitch.

Technical Action

4. Secure rope and chain twitches by rapidly twisting wooden handle.

a. The humane twitch is secured by pressing arms together (like a nutcracker), then winding the attached string around bottom of arms and clipping to side or ventral D-ring of halter.

Rationale/Amplification

4a. To keep twitch in place, the handler must not pull it downward. Think of pushing the twitch into the nose.

4b. The most common reason for twitches coming off at inappropriate times is that the handler pulls down.

4c. If a self-retaining twitch is not needed, the humane twitch does not need to be clipped to the halter.

Technical Action

5. To remove twitch, untwist and then rub horse's nose with the palm of your hand to stimulate circulation.

Rationale/Amplification

5a. Twitches should not remain in place more than 20 minutes without loosening temporarily to facilitate circulation.

Loading Horses in Trailers

Purpose

- Place horse safely in trailer for transport

Complications

- Injury to personnel (crushing, rope burn)
- Trauma to horse (head or legs most commonly)

Equipment

- Halter
- Lead rope
- Protective equipment such as wraps, bell boots, or head bumpers

---

PROCEDURE
Loading Horses in Trailers

Technical Action

---

1. Familiarize yourself with the trailer, noting specifically where the horse is tied, method of closing door, and presence of any emergency exits.

Rationale/Amplification

1a. Many types and models of trailers are available. These can include stock trailers, designed to carry both horses and cattle, 2-horse and 4-horse regular-load trailers, and 2-horse to 4-horse slant-load trailers.

Technical Action

---

2. Apply any protective equipment owner elects to use.

Rationale/Amplification

2a. Protective equipment helps prevent injury to the horse during transit.

2b. Equipment commonly includes leg wraps, bell boots, head bumpers, tail wraps, and blankets.

Technical Action

3. Lead horse onto trailer.

Rationale/Amplification

3a. Do not look at horse when loading. Look toward where you want to go.

3b. *See Figure 7-10.*

3c. If the horse initially refuses to enter, allow it to look at the trailer for 1–2 minutes, then begin clucking to encourage forward motion.

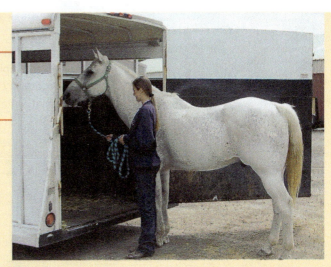

Figure 7-10: Loading a horse in a step-up trailer.

Technical Action

4. Secure butt rope, internal gate, or boom arms, if present. Tie horse using quick-release knot.

Rationale/Amplification

4a. Always use a quick-release knot. This knot can be released with a quick pull on the rope end.

4b. Refer Chapter 6.

Technical Action

5. Secure all exterior doors.

Special Handling Scenarios

Foals

Foals are very delicate and often lack the handling experience of older animals. The handler should take care to assess the individual foal's temperament and degree of training. Make every attempt to keep the mare and foal within sight of each other. In general, the closer the mare is to the foal the better the result of the procedure.

Leading Foals

Technique is dependent upon the amount of prior training.

- If the foal is not halter broken, the best method is to cradle the foal in your arms and push it up against the mare's side.

- If the foal has some experience with the halter, a rump rope can be used. Attach lead rope to the ventral halter D-ring. Use this lead rope as the standard lead rope, then place a second lead rope around foal's rump and back through the halter. Apply more pressure to the butt rope than to the halter lead rope (approximately 70-to-30 percent of pressure). *See Figure 7-11.*
- If the foal is halter broken, use the techniques described for adult horses.

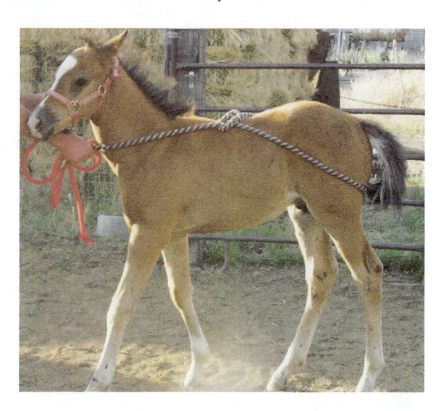

Figure 7-11: Application of a rump rope to a foal.

PROCEDURE
Cradling Foals

Technical Action

1. Place one arm around foal's chest and the other around the rump. Cradle foal in arms.

Rationale/Amplification

1a. Some foals will slump in the handler's arms. If this happens, loosen hold and foal will begin bearing its own weight.

1b. *See Figure 7-12.*

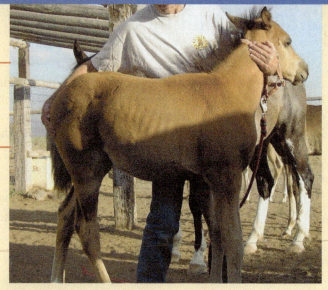

Figure 7-12: Cradling a foal.

Technical Action

2. If a foal is accepting the procedure being performed, shift so that foal is against the side of the mare or wall.

Rationale/Amplification

2a. Mare should also be very calm.

Technical Action

3. With exuberant foals, the handler can grasp the base of the tail.

Rationale/Amplification

3a. Do not pull tail upward or the foal will attempt to sit down.

Stallions

Although most stallions are well behaved, they are much more exuberant and volatile than geldings and mares. In general, they are more likely to strike, bite, or rear. Knowing this, special care should be taken to maintain adequate physical distance between stallions and other horses. A minimum of 20 feet should be maintained between stallions and other animals. This distance is especially important if the stallion must be in the presence of mares in heat (estrus). Novice horseman would be advised to avoid handling stallions until they are very competent handling mares and geldings.

Review Questions

1. What is the reason for most accidents that occur when handling a horse?
2. Describe the proper way to approach a horse.
3. State three rules that should be followed to ensure safety when tying horses.
4. What is the purpose of stock?
5. Describe the proper way to lead a horse.
6. What is the purpose of applying chains?
7. What is the purpose of a twitch?
8. Name some of the protective equipment that can be used when trailering and transporting a horse.
9. Describe the procedure for cradling a foal.
10. What are some of the differences between handling a stallion and handling a mare or gelding?

Bibliography

Fowler, M. (2008). *Restraint and handling of wild and domestic animals* (4th ed.). Ames: Iowa State University Press.

Mackenzie, S. (1998). *Equine safety.* Clifton Park, NY: Thomson Delmar Learning.

McCurnin, D., & Bassert, J. (2005). *Clinical textbook for veterinary technicians* (7th ed.). Philadelphia: W. B. Saunders.

Sirois, M., & Han, C. (2004). *Principles and practice of veterinary technology* (2nd ed.). St. Louis: Mosby.

Chapter 8

Restraint of Cattle

All the really good ideas I ever had came to me when I was milking a cow

—Grant Wood

Objectives

- Identify behavioral characteristics unique to cattle.
- Describe restraint techniques used on cattle.
- Identify various facilities, tools, and equipment used in restraint of cattle.

Key Terms

casting
chute
hot shot
nose ring
sweep tub

Restraint of Cattle

Cattle exhibit marked variance in their response to handling. Dairy cows or 4-H animals accustomed to intimate daily contact tolerate restraint more readily than range cattle. The selection of restraining facilities and equipment directly reflects this variance. Wise handlers will always assess breed, sex, and production use prior to unloading any cattle.

Guidelines for Restraint of Cattle

- Cattle are not typically haltered or led. They are pushed or driven via arm waving and similar measures. Handlers must remember to avoid standing where they want the cattle to go. For example, do not stand in front of the **chute** if you want the cow to exit the chute through the front or head gate. Care should be taken not to overly agitate the cattle when driving them.
- Cattle are very herd oriented. Hence, they are much easier to manipulate as a group than as individuals.
- The fight-flight distance for cattle is approximately 15–20 feet, although this can vary dramatically from breed to breed.
- Always inspect facilities prior to processing. Familiarize yourself with the chute, and ensure that fences are sturdy and gates are locked. Loose cattle are extremely difficult to catch.
- Cattle breeds vary dramatically with regard to temperament. In general, dairy breeds are much more docile than beef breeds. (This statement is true for cows, not bulls.)
- Caution should always be exercised when working with bulls. This is especially relevant when working with dairy bulls.
- Avoid the use of dogs unless the dog is extremely well trained.
- Although most cattle are not overtly aggressive, the following actions can cause damage to personnel.
  1. Butting: Cattle will swing their heads and use them as battering rams. This is especially dangerous in horned animals.
  2. Kicking: The cow kick is usually forward and to the side. Restrained cattle are less likely to kick caudally although this can happen. Cattle almost never kick with both hind limbs simultaneously.
  3. Trampling: Cattle will often run directly over individuals who are blocking their path of escape.
  4. Biting: Cows rarely, if ever, bite.

Dramatic advances have been made in recent years regarding humane handling and processing of cattle. Many individuals have contributed to the welfare of cattle and revolutionized our understanding of bovine restraint. Review of such material is advised.

Processing Facilities

Cattle typically are processed using facilities that have the following equipment. *See Figure 8-1.*

- Chute: This piece of equipment provides immobilization via a head catch and squeezable sides. The side panels can be dropped to examine feet and legs, while the side bars can be dropped to examine the dorsum of the animal.

Head plates with halters can be attached to the head catch to facilitate dehorning. Many models and brands are available. Technicians and veterinarians should familiarize themselves with individual chute operation prior to use. *See Figure 8-1a.*

- Stock: A stock is similar to a chute but does not possess squeezable sides. It is not meant for processing large numbers of cattle and should never be used with fractious animals.
- Alley way: The alley way is the narrow passage area that prevents cattle from turning around as they approach the chute. Poles or back gates can be used to prevent cattle from backing up in the alley way as they approach the chute. *See Figure 8-1b.*

Figure 8-1a: Cow in a squeeze chute.

Figure 8-1b: Alley way.

Figure 8-1c: Sweep tub.

- Sweep tub: **Sweep tubs** permit a small group of cattle to be squeezed together to facilitate passage in the alley way. Minimizing excess room space decreases the opportunity for cattle to turn around and refuse to enter the alley way. *See Figure 8-1c*.
- Palpation cage: These gates are placed directly behind the chute. They facilitate entrance into the tailgate of the chute by prohibiting cattle in the alley way from approaching the chute. *See Figure 8-1d*.

Figure 8-1d: Chute with palpation cage.

Operating Chutes

Purpose

- Provide the most effective method of restraint for cattle

Complications

- Injury to cattle
- Injury to personnel

Equipment

- Chute

PROCEDURE
Operating Chutes

Technical Action

1. Ready chute for operation by opening head catch, releasing squeeze, and opening tailgate.

Rationale/Amplification

1a. Familiarize yourself with the individual chute operation before running cattle. There are many models available, each with slight variations in operation.

Technical Action

2. Allow cow to enter.

Technical Action

3. Catch head.

Rationale/Amplification

3a. Experienced chute operators will often close the tailgate before catching the head. This prevents the cow from backing out of the chute before her head is caught. Slow operators will miss cattle using this technique.

Technical Action

4. Close tailgate.

Technical Action

5. Apply squeeze.

Rationale/Amplification

5a. Chutes can accommodate various sizes of animals by manually moving the side panels inward or outward. If the cow can't be squeezed sufficiently, check the position of the side panel. If needed, move the panel inward by adjusting the pins located on the ventral corners.

Technical Action

6. To release cow, release squeeze and then release head catch.

Rationale/Amplification

6a. If a dehorning plate is attached, the cow will need to be released from the side of the chute, instead of through the head catch.

Haltering

Purpose
- Provide restraint of head
- Permit leading of cattle that have been halter broken

Complication
- Trauma to personnel

Equipment
- Cow halter

PROCEDURE
Haltering Cattle

Technical Action

1. Place cow in chute or stock.

Rationale/Amplification

1a. If cow has been broken to a halter, it can be applied in the stall or pen.

Technical Action

2. Place crown piece of halter over ears, then slip nose through nosepiece. Adjust halter such that nose band crosses over bridge of nose halfway between the nostrils and eyes.

Rationale/Amplification

2a. Cattle halters typically are made of a single piece of rope that forms both the halter and lead rope. The halter is adjustable and should be fitted individually to each cow.

2b. The adjustable portion of the nose band should always go under the chin, not across the bridge of the nose. The standing end or lead rope portion should be on the left side of the cow.

2c. *See Figure 8-2.*

Figure 8-2: Haltering a cow.

Tailing-Up Cattle

Purpose

 • Provide restraint for examination or minor surgical procedures

Complications

- Injury to coccygeal vertebrae
- Injury to personnel

Equipment

- None

PROCEDURE
Tailing-up Cattle

Technical Action

1. Place cow in chute or stock and catch head.

Rationale/Amplification

1a. *See Figure 8-3.*

Figure 8-3: Tailing-up a cow.

Technical Action

2. Stand caudal to cow and use both hands to grasp the base of the tail, 5–6 inches from the anus. Push the tail straight over the dorsum. Do not permit tail to curl or to shift left to right.

Rationale/Amplification

2a. Movement or deviation of the tail dramatically decreases the effectiveness of tailing-up. A cow is very likely to kick if inappropriate technique is used.

2b. This technique should not be confused with tail twisting, which involves bending the tail to make a cow move forward.

2c. Grasping the tail base decreases the likelihood of damaging coccygeal vertebrae.

Technical Action

3. Maintain pressure until directed to release.

Rationale/Amplification

3a. This hold can result in arm fatigue.

Casting Cattle

Purpose

- Place cow in lateral or sternal recumbency

Complications

- Tissue trauma
- Cow resists recumbency

Equipment

- Forty feet of heavy cotton rope
- Halter

PROCEDURE
Casting Cattle

Technical Action

1. Place halter on cow.

Technical Action

2. Ensure that ground surface is suitable for recumbency.

Technical Action

3. Secure cow to fence post. Tie knot low to the ground.

Rationale/Amplification

3a. The cow must be secured to prevent it from backing up.

3b. Knot must be low to prevent cow from hanging in recumbency.

Technical Action

4. Locate midpoint of rope and place over dorsum of neck.

Technical Action

5. Run rope medial to forelimbs, crisscross over back, then run medial to hind limbs.

Rationale/Amplification

5a. *See Figure 8-4.*

5b. The Burley method of **casting** is superior to the half-hitch method in that it does not place pressure on the trachea, penis, or mammary veins. The rope ends can also be used to tie the hind limbs in a flexed position, thus eliminating the need for an assistant to hold the ropes.

5c. If the cow kicks, gently toss the rope between the hind limbs instead of passing it through with your hand.

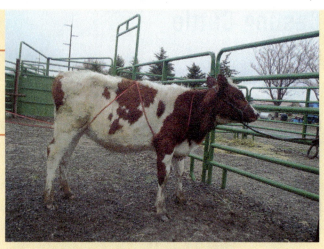

Figure 8-4: Rope placement for casting a cow.

Technical Action

6. Standing caudal to cow, pull on rope ends until cow drops into recumbency.

Rationale/Amplification

6a. This technique is often more successful if two people work together, with one holding each rope end.

Flanking

Purpose

- Place calf or goat in lateral recumbency

Complications

- Tissue injury to animal
- Back pain in restraining personnel

Equipment

- None

PROCEDURE
Flanking Calves

Technical Action

1. Bring animal to an area suitable for recumbency.

Rationale/Amplification

1a. Animals that weigh over 200 lbs should not be flanked.

Technical Action

2. Stand on left side of animal.

Rationale/Amplification

2a. *See Figure 8-5.*

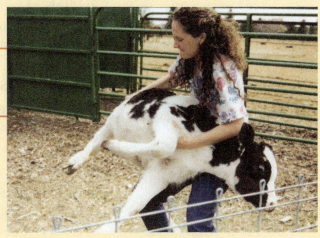

Figure 8-5: Flanking a calf to lay it on the ground.

Technical Action

3. Place left arm over neck and grasp ventral neck area. Reach over animal and use right hand to grasp right flank area.

Technical Action

4. Push right knee into animal's flank. As calf jumps forward, use this momentum to lift animal off its feet.

Rationale/Amplification

4a. It is critical to use the animal's own momentum, otherwise you will not be able to lift the calf.

Technical Action

5. Bend knee to push the left side underneath the calf, and quickly press animal down onto its left side.

Rationale/Amplification

5a. Maintain position by placing your left knee over the animal's neck.

5b. Holding the cannon bone of the down forelimb will inhibit the calf from rising.

Securing Cattle Feet for Examination

Purpose

- Provide immobilization of foot to facilitate examination of hoof, application of bandage, hoof trimming, and other procedures

Complications

- Cow becomes recumbent in chute
- Trauma to personnel secondary to kicking

Equipment

- Cotton rope 15–30 feet
- Chute or stock

PROCEDURE
Securing Feet for Examination

Technical Action

1. Place cow in chute or stock.

Rationale/Amplification

1a. Tilt tables can be used to examine feet and are far superior to tying up the feet.

Technical Action

2. Remove side panel.

Technical Action

3. Tie square knot or clove hitch around cannon area of leg. Apply half hitch distal to square knot.

Rationale/Amplification

3a. Take care to avoid being kicked during this maneuver.

Technical Action

4. Slip running end of rope over beam, and pull to lift leg off ground.

Rationale/Amplification

4a. At this point, the leg is off the ground and can be examined visually but the cow can still kick.

Technical Action

5. Apply a second half hitch over fetlock, and pull limb tight against the corner post of the stock or chute.

Rationale/Amplification

5a. *See Figure 8-6.*

Figure 8-6: Securing a hind foot for examination or treatment.

Technical Action

6. Apply third or fourth half hitches as needed to minimize motion of feet.

Hot Shot

Encouraging cattle to move through processing facilities often requires ancillary equipment such as the **hot shot**, prod, or paddle (*Table 8-1, Figure 8-7*). This equipment offers varying degrees of encouragement to the livestock and should be used with care and compassion.

Table 8-1 Equipment Description

Equipment Name	Composition or Appearance	Comments
Hot shot	Battery-powered electronic device	Device causes intense pain. Use with caution. Avoid use if possible.
Prod	Graphite rod	—
Paddle	Plastic and looks similar to a boat paddle	They are typically filled with beads to make a rattling sound.
Pole	Pole that is attached to the nose of a bull ring	A pole prevents the bull from moving toward a handler.

Figure 8-7: Cow processing equipment including (a) a prod (b) a hot spot, and (c) a paddle.

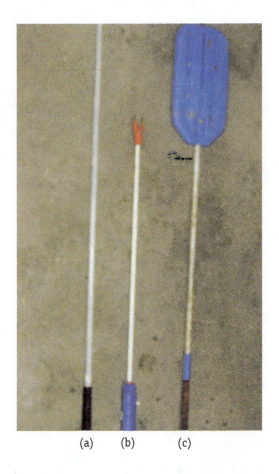

(a)   (b)   (c)

Purpose

- Equipment used to stimulate cattle to move through a processing area or stop movement toward a handler

Complication

- Pain and distress

PROCEDURE
Use of a Hot Shot

Technical Action

1. Cow must be in a contained alley way or chute area.

Rationale/Amplification

1a. Never apply a hot shot to an animal in an open pen or pasture.

Technical Action

2. Remove safety clip, if present.

Rationale/Amplification

2a. The safety clip prevents accidental discharge of the hot shot.

Technical Action

3. Confirm that battery is operational by depressing button. A distinct buzzing sound should be heard.

Technical Action

4. Touch cow with hot shot.

Technical Action

5. Before putting hot shot down, touch the end against a metal surface. This will discharge the residual electricity.

Rationale/Amplification

5a. Replace safety clip, if present.

5b. Hot shots should be used only as a last resort. They cause extreme pain. Anyone who uses a hot shot frequently should probably experience the procedure; it will help one to develop empathy.

⌒ Nose Tongs

Purpose

- Restraint device used to control head

Complications

- Damage to nasal septum
- Personnel trauma
- Pain and discomfort to animal

Equipment

- Tongs
- Lead rope
- Halter

PROCEDURE
Placing Nose Tongs

Technical Action

1. Place cow in chute or stock and catch head.

Technical Action

2. Apply halter if available.

Rationale/Amplification

2a. Many times, a halter or head plate is not available, which is why the tongs are used.

Technical Action

3. If halter is not used, restrain by grasping head and pin between left arm and thigh.

Rationale/Amplification

3a. Restraint of the head is difficult. Cattle will often swing their heads about in an attempt to avoid tongs placement.

Technical Action

4. Place tongs using a rotating motion. Insert one side of the tongs, then rotate across the nostril, and place other on opposite side.

Rationale/Amplification

4a. Use only tongs that have a space between the balls. Tongs which leave no room for the nasal septum are considered inhumane.

4b. Before placement, always examine balls to ensure that surfaces are smooth.

Technical Action

5. Apply tension to maintain tongs in position. A lead rope can be attached to the end of most tongs.

Rationale/Amplification

5a. Never leave an animal unattended when nose tongs are in place.

5b. *See Figure 8-8.*

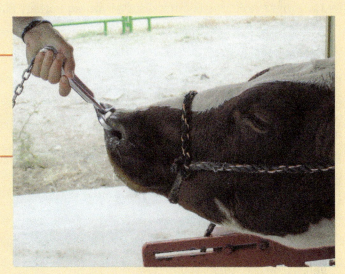

Figure 8-8: Application of nose tongs.

Nose Rings

Purpose
- A permanent restraint device used to control head of bulls

Complication
- Torn nasal septum

Equipment
- Self-piercing **nose ring** with screw and Allen wrench
- Cloth rag

PROCEDURE
Placing a Nose Ring

Technical Action

1. Place bull in chute.

Rational/Amplification

1a. Proper restraint is critical for attaining correct placement.

Technical Action

2. Place halter or secure head in head plate.

Rational/Amplification

2a. Head plates are much preferred over halters, because they immobilize the head better.

Technical Action

3. Select ring of the appropriate size.

Rational/Amplification

3a. Rings typically are placed when bulls are 1–2 years of age. Do not use large rings on these small bulls, because they will enlarge the nasal septum hole inappropriately.

3b. Ring sizes are small: 2.5 inches, medium: 3 inches, and large: 3.5 inches.

3c. Rings that are too large or placed too close to the end of the septum will hang out and catch on objects. A torn nasal septum can be catastrophic to a bull's breeding career.

Technical Action

4. Apply topical anesthetic or inject 5.0 cc lidocaine into the septum. Injection should be made 1–2 inches from the end of septum.

Rational/Amplification

4a. Nose ring placement is extremely painful for the animal. Postsurgical analgesics should be used.

Technical Action

5. Open ring and use sharpened end to pierce through nasal septum. Once through, use rag to wipe screw area of the ring free of blood.

Technical Action

6. Place screw in ring and tighten using Allen wrench.

Rational/Amplification

6a. Be careful with the screw. They are easy to drop and lose.

Review Questions

1. Explain some strategies for herding or moving cattle.
2. Describe behavior of cattle that can cause injury to the handler.
3. Identify the primary difference between a chute and a stock.
4. List the standard facilities used to process cattle.
5. What is the purpose of tailing up cattle?
6. Describe the method used to flank a calf.
7. Diagram the placement of the ropes used for casting cattle.
8. Describe assistive devices that can be used to move cattle. What are the advantages and disadvantages of such devices?
9. Compare and contrast nose rings and nose tongs.
10. Explain behaviors in cattle that a handler should be aware of during a restraint procedure.

Bibliography

Fowler, M. (2008). *Restraint and handling of wild and domestic animals* (4th ed.). Ames: Iowa State University Press.

McCurnin, D., & Bassert, J. (2005). *Clinical textbook for veterinary technicians* (6th ed.). Philadelphia: W. B. Saunders.

Noordsy J. (2006). *Food animal surgery* (2nd ed.). Lenexa, KS: Veterinary Medicine Publishing.

Sirois, M., & Han, C. (2004). *Principles and practice of veterinary technology* (2nd ed.). St. Louis: Mosby.

Courtesy of Image Source

Chapter 9

Restraint of the Goat

Objectives

- Identify behavioral characteristics unique to the goat.
- Describe restraint techniques used on the goat.
- Identify various facilities, tools, and equipment used in restraint of the goat.

Key Term

stanchion

Don't approach a goat from the front, a horse from the back, or a fool from any side.

—Yiddish proverb

Restraint of the Goat

Goats are unique in their dual role as both production and pet animals. Discerning individual animal status is best accomplished through consultation and observation of the animal with the owner. Goats fulfilling pet roles should be referred to by name. Interactions with this category of goat should parallel relationships to pets such as dogs. Alternatively, interactions with production goats will mirror interactions with cattle.

Guidelines for Restraint of the Goat

- Goats are gregarious, fun-loving animals that do not possess a strong herd instinct. They typically respond well to gentle handling and willingly accompany handlers away from the herd. Goats are not typically haltered but are led using a collar around the neck.
- Distraught or angry goats will vocalize and stamp their forefeet. Although they do not usually charge, extremely stressed goats may attempt to jump over a handler. This typically results in the feet of the goat being placed in the center of the handler's chest as the animal attempts to spring over.
- Horned goats may attempt to butt the handler. Goats do not bite or kick.
- Scent glands contribute to the unpleasant odor of intact males. This foul smell is especially noticeable during breeding season. Bucks will also urinate on their beards and forelimbs to increase their attractiveness to the does. Wise handlers do not stand unaware in front of bucks during the breeding season.

Collaring and Leading Goats

Purpose

- General restraint

Complications

- None

Equipment

- Collar
- Lead rope

PROCEDURE
Collaring and Leading Goats

Technical Action

1. Place collar on neck, leaving 2 inches of space between collar and neck.

Rationale/Amplification

1a. Most goats receiving veterinary care are considered companion animals. As such, they are typically handled in a manner consistent with that of other large animals of companion status (e.g., horse, llama). They are not handled in the same way as are production animals (e.g., cow, pig). This distinction is very important to goat owners.

Technical Action

2. Attach lead rope and lead animal from left side.

Rationale/Amplification

2a. Most goats are accustomed to handling and will follow readily.

2b. If a lead rope is not available, it is acceptable to grasp the collar directly.

Stanchion

Purpose

- Restrain for hoof trim, milking, artificial insemination, or examination

Complication

- Goat jumps off **stanchion**

Equipment

- Stanchion
- Collar

PROCEDURE
Placing Goat in Stanchion

Technical Action

1. Lead goat to stanchion using collar.

Rationale/Amplification

1a. A stanchion is essentially a raised platform with a head catch. It is not suitable for fractious animals.

Technical Action

2. Encourage goat to jump onto platform by clucking and lifting up on collar. Secure head in head catch.

Rationale/Amplification

2a. Many stanchions have attached grain boxes. Placing a handful of grain in the box rewards the goat for good behavior.

2b. Milk goats readily accept stanchions; however, production and pet goats may require encouragement. *See Figure 9-1a.*

2c. If a stanchion is unavailable, the animal can be straddled or held by the collar. *See Figures 9-1b and 9-1c.*

Figure 9-1a: Goat secured in stanchion.

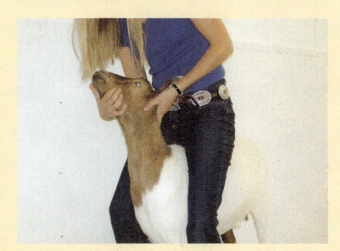

Figure 9-1b: Straddling goat to facilitate jugular access.

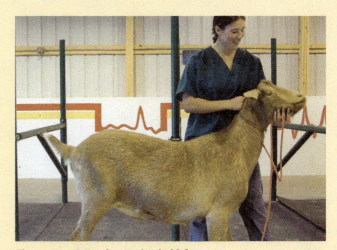

Figure 9-1c: Lateral restraint hold for jugular access.

Chapter Review

1. Describe the behaviors of goats a restrainer may need to be alert to.
2. Name the equipment used to restrain goats for milking.

Bibliography

Fowler, M. (2008). *Restraint and handling of wild and domestic animals* (4th ed.). Ames: Iowa State University Press.

McCurnin, D., & Bassert, J. (2005). *Clinical textbook for veterinary technicians* (6th ed.). Philadelphia: W. B. Saunders.

Noordsy J. (2006). *Food animal surgery* (2nd ed.). Lenexa, KS: Veterinary Medicine Publishing.

Sirois, M., & Han, C. (2004). *Principles and practice of veterinary technology* (2nd ed.). St. Louis: Mosby.

Courtesy of Image Source

Chapter 10

Restraint of the Pig

Objectives

- Identify behavioral characteristics unique to pigs.
- Describe restraint techniques used on pigs.
- Identify various facilities, tools, and equipment used in restraint of pigs.

Key Terms

pig board
snout snare

A dog looks up to you, a cat looks down upon you, but a pig will look you right in the eye.

—Winston Churchill

Restraint of Pigs

Intelligent and independent animals such as pigs do not appreciate restraint, and they never hesitate to voice their displeasure when subjected to this insult. Use of appropriate processing facilities can help to minimize this disruption; however, many pig owners do not have these facilities available.

Guidelines for Restraint of the Pig

- Pigs are independent animals with little herd instinct. Although they do not seek comfort from a herd, pigs show great concern when fellow pigs are distressed. Thus, handlers should take care when working within a confined group.
- Most pigs are not aggressive. When stressed, however, pigs can inflict extensive damage using their teeth. This is especially true with boars, which have elongated canine teeth (tusks).
- Sows with litters are very protective. Care should be taken to confine the sow when handling the piglets.

Pig Boards

Purpose

- General restraint for examination or moving pigs from one area to another

Complications

- Pig escapes under board
- Handler knocked over

Equipment

- **Pig board** (3-foot by 4-foot piece of plywood with hand holds)

PROCEDURE
Using Pig Board

Technical Action

1. Hold board parallel to pig.

Rationale/Amplification

1a. Keep board close to the ground.

Technical Action

2. Use board to push pig gently toward corner of enclosure.

Castration Restraint

Purpose

- Restrain pig for **castration**

Complication

- Handler bitten on leg or knocked to ground

Equipment

- None

PROCEDURE
Holding Pig in Castration Position

Technical Action

1. Grasp both hind limbs proximal to hocks.

Rationale/Amplification

1a. This hold is also called the pig handstand.

1b. Use of suitable ear plugs for the protection of the handler and surgeon is advised.

Technical Action

2. Position pig such that handler is standing over pig, with one leg on either side. The pig's head should be between the handler's legs, facing the opposite direction.

Rationale/Amplification

2a. *See Figure 10-1.*

2b. Do not twist the hind limbs, because this can dislocate the hips.

2c. Small pigs can be lifted so that forefeet do not touch the ground.

Figure 10-1: Pig castration restraint.

Snout Snare

Purpose

- Used as a restraint device on larger pigs for sample collection or other clinical procedures

Complications

- Damage to entire snout as result of tourniquet effect
- Damage to nasal cartilage

Equipment

- **Snout snare**

PROCEDURE
Applying a Snout Snare

Technical Action

1. Approach pig from left or right side caudal to head.

Rationale/Amplification

1a. *See Figure 10-2.*

Figure 10-2: Application of a snout (hog) snare.

Technical Action

2. Place loop over snout, working caudally until loop is caudal to canines.

Rationale/Amplification

2a. Snares that are placed too far cranially will cause pain from pressure on nasal cartilage.

Technical Action

3. Tighten cable and apply pressure toward the pig.

Rationale/Amplification

3a. Pigs will pull backward against the snare and squeal loudly.

3b. Do not use snares for longer than 20-minute intervals, because the tight cable will act as a tourniquet.

3c. Most pigs will permit a snare application 2 or 3 times. After this, they will use evasive maneuvers, making it impossible to apply the snare.

Technical Action

4. To remove snare, step toward pig, loosen cable, and remove from mouth.

Review Questions

1. Explain behaviors of pigs that restrainers may need to be alert to.
2. Explain the use of a pig board.
3. State two complications associated with the use of a pig snout snare.

Bibliography

Fowler, M. (2008). *Restraint and handling of wild and domestic animals* (4th ed.). Ames: Iowa State University Press.

McCurnin, D., & Bassert, J. (2005). *Clinical textbook for veterinary technicians* (6th ed.). Philadelphia: W. B. Saunders.

Noordsy J. (2006). *Food animal surgery* (2nd ed.). Lenexa, KS: Veterinary Medicine Publishing.

Sirois, M., & Han, C. (2004). *Principles and practice of veterinary technology* (2nd ed.). St. Louis: Mosby.

Delmar/Cengage Learning

Chapter 11

Restraint of the Llama

Objectives

- Identify behavioral characteristics unique to llamas.
- Describe restraint techniques used on llamas.
- Identify various facilities, tools, and equipment used in restraint of llamas.

The llama is a wooly sort of fleecy hairy goat, with an indolent expression and an undulating throat; like an unsuccessful literary man.

—Hilaire Belloc

Restraint of the Llama

A gradual rise in llama popularity has brought these animals out of the exotic designation into a routine large animal arena. Technicians working in mixed or large animal practices will most likely encounter llamas with some frequency. Unique in their own right, llamas are restrained using techniques common to both horses and cattle. For example, llamas are haltered routinely and led, yet they are also restrained with head catches. Ultimately, familiarity with handling will be the most influential factor in determining the restraint techniques used on each animal.

Guidelines for Restraint of the Llama

- Llamas are social and curious animals. They vary dramatically in their response to restraint, and this is most likely the result of prior handling experiences. Some llamas are extremely well trained, while others may never have been touched. Therefore, it is advisable to talk with the owner to determine proper restraint protocol.
- Although most llamas are very docile, they can cause harm in a number of ways. These include biting (especially dangerous in intact males with canine teeth), spitting, regurgitation, and kicking. Most llamas will kick like a cow does, but occasionally they will kick directly behind with one foot.

Haltering and Leading

Purpose

- Secure for examination or to move animal

Complications

- Unable to catch llama
- Spitting or biting

Equipment

- Halter and lead rope

PROCEDURE
Haltering and Leading the Llama

Technical Action

1. Approach llama quietly from the left side. Do not make direct eye contact.

Rationale/Amplification

1a. Llamas are accustomed to being handled from the left side.

1b. Direct eye contact can be interpreted as confrontational behavior.

Technical Action

2. Place lead rope around neck.

Technical Action

3. Place nose band over nose, and secure crown piece behind ears.

Rationale/Amplification

3a. Limit contact with ears if possible.

Technical Action

4. Lead llama from the left side.

Rationale/Amplification

4a. Optimally, the handler remains between the llama's head and shoulder while leading.

4b. *See Figure 11-1.*

Figure 11-1: Leading a llama.

Stock

Purpose
- Inhibit llama movement through bodily confinement

Complication
- Lying down in the stock

Equipment
- Stock
- Halter and lead rope

PROCEDURE
Placing Llama in a Stock

Technical Action

1. Halter the llama.

Technical Action

2. Open chute or stock head catch, tailgate, and release squeeze.

Rationale/Amplification

2a. Llamas can be placed in chutes or stocks.

2b. Stocks designed specifically for llamas are available. Most veterinary clinics, however, will use stocks designed for horses and cattle.

Technical Action

3. Walk llama into chute or stock.

Rationale/Amplification

3a. If the handler must walk the llama through a chute, two people are required. One will operate the chute, and the other will handle the llama. Stocks can be managed by an individual.

Technical Action

4. Secure head catch.

Rationale/Amplification

4a. *See Figure 11-2.*

Figure 11-2: Llama stock.

Technical Action

5. Close tail gate.

Technical Action

6. Apply squeeze if using chute.

Rationale/Amplification

6a. Do not apply a snug squeeze, because many llamas will lie down in response to squeezing.

Review Questions

1. Describe llama behaviors a restrainer may need to be alert to.
2. Identify the animal species most likely to regurgitate when stressed.
3. Explain the proper way to lead a llama.

Bibliography

Fowler, M. (1998). *Medicine and surgery of South American camelids* (2nd ed.). Ames: Iowa State University Press.

Fowler, M. (2008). *Restraint and handling of wild and domestic animals* (4th ed.). Ames: Iowa State University Press.

McCurnin, D., & Bassert, J. (2005). *Clinical textbook for veterinary technicians* (6th ed.). Philadelphia: W. B. Saunders.

Sirois, M., & Han, C. (2004). *Principles and practice of veterinary technology* (2nd ed.). St. Louis: Mosby.

Glossary

bight – Fold of rope.

brachycephalic – Site of venipuncture in dog.

casting – The use of ropes to lay down cattle.

castration – The removal of both testicles.

cat bag – Restraint device for a cat.

catheterization – Insertion of a tubular device into a duct, blood vessel, hollow organ, or body cavity for injecting or withdrawing fluids for diagnostic or therapeutic purposes.

caudal – A directional term or reference point describing toward the hind end.

chute – An adjustable restraint device used for cows. Similar in appearance to a stock.

cranial – A directional term or reference point describing toward the head.

cystocentesis – Procedure to withdraw urine sample.

dyspnea – Difficulty breathing.

halter – A restraint device that fits around the head and is used to lead or secure an animal.

hondo – Loop through which the rope passes in a lariat.

hot shot – A probe with an electrical shock used for encouraging animals to move.

jugular – The blood vessel (vein) in the neck which drains blood from the head and conveys it toward the heart.

lariat – Coated, braided rope designed for catching livestock out in the field, pasture, or pen. Used by cowboys on the ranch or in the rodeo.

lateral – A directional term describing toward the left or right side of the body, away from the midline.

loop – An ovoid or circular shape formed by crossing the ends of a rope.

medial – A directional term describing toward the middle or midline of the body.

muzzle – Restraint device to prevent biting.

noose leash – Restraint device for control and handling of dog.

nose ring – A ring inserted through the nasal septum for purposes of restraint.

pig board – A board used to restrain a pig.

proptosis – Bulging of eyes.

rabies pole – Restraint device to control dog.

recumbency – In a lying down position. Lateral recumbency would be lying on one's side, and dorsal recumbency would be lying on one's back.

restraint – Forcible confinement.

scruffing – Restraint of animal by pinching for skin at base of neck.

sisal – Fiber obtained from an agave plant used to make rope.

snout snare – A restraint device that goes around a pig's snout.

stock – A nonadjustable restraint device consisting of vertical pillars arranged in a rectangular shape connected by horizontal bars. Used to keep livestock restrained in a standing position.

stanchion – Restraint device used for milking goat.

sweep tub – A round pen in which the gate pushes forward, allowing for cows to be crowded into an alleyway.

tachypnea – An abnormally fast rate of breathing.

tailing-up – Grabbing the base of a cow's tail and elevating it vertically to restrain the animal.

tensile strength – Amount of load or stress a given substance can bear without breaking, when applied lengthwise.

twitch – A restraint device used on the horse's nose.

venipuncture – Technique used to draw blood from a vein for diagnostic purposes or treatment.

ventral – Directed toward or situated on the abdominal surface. The opposite of dorsal.

Index

Note: Page numbers followed by f indicate figures; those followed by t indicate tables.